Mechanism design

Mechanism design

An introductory text

S. MOLIAN

CAMBRIDGE UNIVERSITY PRESS

Cambridge

London   New York   New Rochelle

Melbourne   Sydney

Published by the Press Syndicate of the University of Cambridge
The Pitt Building, Trumpington Street, Cambridge CB2 1RP
32 East 57th Street, New York, NY 10022, USA
296 Beaconsfield Parade, Middle Park, Melbourne 3206, Australia

First published 1982

Printed in Great Britain at the University Press, Cambridge

Library of Congress catalogue card number: 81-15552

British Library catalogue in publication data
Molian, S.
Mechanism design.
1. Machinery − Design
I. Title
621.8'15   TJ230
ISBN 0 521 23193 0 hard covers
ISBN 0 521 29863 6 paperback

MU

CONTENTS

PREFACE

This book is intended for machine designers and students of machine design. It describes some of the simpler (and often more useful) methods of analysing and synthesising mechanisms, which can be carried out on the drawing board or, in some cases, fairly easily on a computer or even a calculator. I have avoided difficult techniques altogether, and omitted proofs of even the easy ones where those proofs are difficult. But at the end of the book there are some suggestions for further reading in case the reader wants to know more. The book is mainly about linkages and gear-trains. I include nothing on cams because elementary methods of laying out a cam are properly part of a technical drawing course and advanced methods, if they are to be used for any sensible purpose, involve difficult considerations beyond the level of what is intended as an elementary book.

Some exercises are provided, in which the design of a mechanism is intended to mean determining those dimensions that settle the kinematics, such as the centre-distances on a link or the number of teeth on a gear. If the book is used in a more general design course these exercises can, where appropriate, be extended to include detail design. I have made some attempt at realism by including exercises that require practical judgment as well as graphical or analytical skill, and to which there are no single correct answers. Most of the benefit of this type of exercise is found when students compare and discuss their results.

Most of the techniques described here are well known to experts in this field and can be found in advanced textbooks. But the constraint method of kinematic analysis given in Chapter 11, and the theory of differential mechanisms given in Chapter 13, have previously appeared only in technical journals.

Most of the material in this book has been used as lecture notes at the Cranfield Institute of Technology, and I am indebted to many of the Institute's students for corrections and improvements which I have included while putting the work in its present form. I am also indebted to the staff of Cambridge University Press; and to Mrs Julie Saunders, who prepared the illustrations.

1 MAINLY ABOUT LINKAGES

1.1 Introduction

Modern studies in mechanism differ considerably from the traditional theory of machines. The difference is partly one of content — vibration and control theory have developed into separate specialities, leaving the mechanisms specialist free to concentrate on such topics as cams, gears and linkages. But the most important difference is a matter of emphasis rather than choice of topics. It is perhaps best explained by considering the kind of examination question that the theory of machines equips a candidate to answer.

Such a question might take the following form: 'The mechanism shown in Fig. 1.1 is used in a certain experimental engine. Find the angular velocity of the larger gear-wheel when the mechanism is in the position shown, if the linear velocity of the piston at that instant is 100 feet per second.' The theory of machines, as traditionally studied, equips us to solve problems of which this example is typical – to find velocities, accelerations and forces in a *given* mechanism.

That is a useful skill, for we must calculate the forces in a mechanism in order to complete the detail design if the mechanism runs at high speeds or is heavily loaded. But given mechanisms exist only in examination papers. In practice we must choose or devise the form of the mechanism — cams, gears, or links — and the way these components are connected together; and we must find the main dimensions that determine the relative motions in the mechanism. How is the

Fig. 1.1. Geared 5-bar linkage used in an experimental engine.

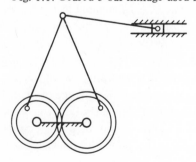

designer to arrive at the type of mechanism shown in Fig. 1.1? How many alternative types are there that will do the same job, and is there any systematic way of finding all of them so that we can make a best choice? Having chosen a type, how are we to determine the main dimensions, in a real design problem where there is no examiner to give them to us?

The modern theory of mechanism sets out to solve these problems by means of the following steps:

- (a)   Enumerating and classifying all possible mechanisms in a definite order;
- (b)   Selecting from amongst the mechanisms discovered (by step (a)) all those that will meet a given specification;
- (c)   Finding the main dimensions that will provide a specified transmission of motion between input and output, graphically or by calculation.

These objectives are clearly of practical importance and there has been a fair measure of success in achieving them. That success is a result of going about the work in a scientific way instead of by trial-and-error. As in any science, much of the work seems at first to be somewhat remote from the day-to-day problems of engineering practice. But practical applications soon become apparent, showing that these methods are of value in actual design work.

The most useful results of recent research on mechanisms are concerned with linkages. These are mechanisms in which the components are connected by pivoted or sliding joints. Linkages using only pivoted joints are of particular interest, since they are the cheapest, simplest and most robust of all mechanisms. They can transmit heavier loads, and run at higher speeds, than other types. They are worth studying, not only for their own sake, but also as an introduction to mechanisms in general.

1.2   Essential features of a linkage

The essential features of a linkage are often disguised by constructional details. The needle-bar drive shown in Fig. 1.2 is a good example of a disguised linkage. Its action is more easily understood if we use a conventional diagram which will now be explained.

Consider the 4-bar linkage shown in Fig. 1.3(a). Suppose this mechanism is used to connect the shafts 1 and 2; it transmits rotary motion with input angle θ and output angle ϕ. The behaviour of the mechanism is not altered if one of the links has a different shape, as in Fig. 1.3(b), provided the centre-distances between the pivots remain the same. Nor does it matter if we enlarge one of the pivots, as in Fig. 1.3(c); in fact we can continue this enlargement until one pivot encloses another, obtaining an eccentric as in Fig. 1.3(d). One of the links can be enlarged until it becomes a complete machine-frame (Fig. 1.3(e)). Despite any of these changes the linkage will transmit motion in the same way provided the four

Fig. 1.2. Needle-bar drive used in a textile machine. A complex linkage such as this is more easily understood if it is re-drawn as in Fig. 1.4.

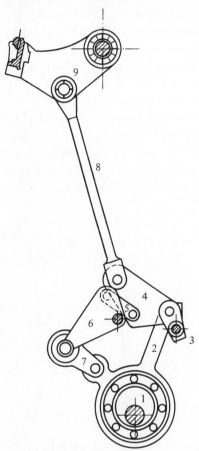

centre-distances (called link-lengths) remain the same. Hence it can be represented by a simple line diagram, as in Fig. 1.3(*f*). In this diagram we shade one of the links to indicate that it is the fixed link, or frame. It is not the lengths of the links (important as they are) that make this mechanism a 4-bar linkage rather than some other type, but simply the fact that there are four links connected together by pivots to form a closed chain. The mechanism shown in Fig. 1.3(*g*), for example, is clearly of the same type.

We can now re-draw Fig. 1.2 by placing a piece of tracing paper over it and pricking through to obtain the pivot-centres. We thus obtain the simpler representation shown in Fig. 1.4(*a*), from which it is much easier to see which of the links are connected together. Two further symbols are introduced in this figure. Where three (or more) links share the same pivot, the small circle representing a pivot is replaced by two concentric circles. Since, in this example, the fixed link overlaps other links we omit it and indicate by local shading that some of the pivots are fixed. It will appear later on that this representation of the fixed link is not as convenient as that used in Fig. 1.4(*b*), which shows a mechanism of the same type but with different link dimensions. Corresponding links in these two diagrams have been given the same numbers to make it easier to see that each mechanism consists of the same kinds of links connected together in the same order, despite their very different appearance.

The importance of specifying which link is fixed is illustrated by the slider-crank mechanism shown in Fig. 1.5(*a*). The simple engine mechanism is an example of this type. As indicated in the figure, the mechanism will behave in the same way whether the slider surrounds the guide or lies inside it – in other words we could, in principle, interchange piston and cylinder. A symbolic representation

Fig. 1.3. The 4-bar linkage. The essential features of any such linkage can be shown by simple line diagrams, as in (*f*) and (*g*), showing only the lines joining the pivot-centres. The fixed link (frame) is indicated by shading.

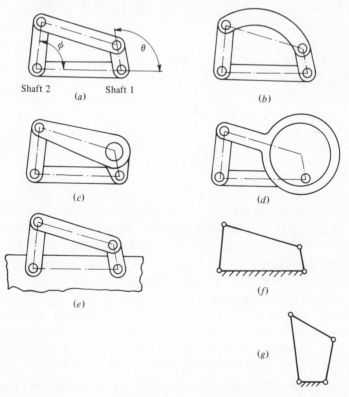

Fig. 1.4. The needle-bar drive (Fig. 1.2) is shown in (*a*) in a conventional diagram, maintaining the same bearing-centres. A linkage of the same type but with different proportions is shown in (*b*). Corresponding links in these diagrams (and in Fig. 1.2) bear the same numbers. Note the two different ways of indicating the frame.

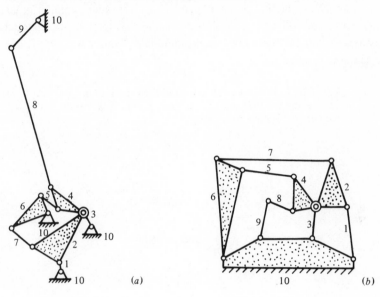

(a)　　　　　(b)

Fig. 1.5. Practical forms and conventional diagram of the slider–crank mechanism with fixed slideways.

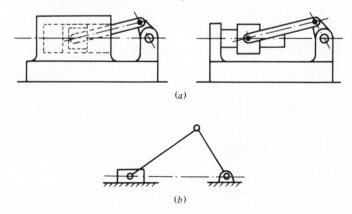

(a)

(b)

of this mechanism is shown in Fig. 1.5(*b*). Now suppose we fix the crank instead of the cylinder, as in Fig. 1.6(*a*). We obtain the rotary engine (*b*) and the quick-return mechanism (*c*). Replacing the slider by a pin, we also obtain the Geneva mechanism (*d*). Fixing the connecting-rod and interchanging piston and cylinder, we obtain the oscillating-cylinder engine (Fig. 1.7), once common in marine engineering and still used as a toy steam engine. Fixing the slide and interchanging piston and cylinder, we obtain an uncommon mechanism sometimes used as a hand-pump (Fig. 1.8).

Fig. 1.6. Slider–crank mechanisms with fixed crank, giving the rotary engine and a quick-return mechanism. By replacing the slider with a pin we get the Geneva mechanism.

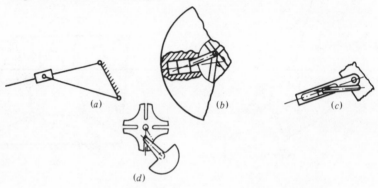

Fig. 1.7. Slider–crank mechanism with fixed connecting-rod. Piston and cylinder have been interchanged (cf. Fig. 1.5) to give a practical form, the oscillating-cylinder engine.

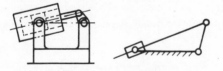

Fig. 1.8. Slider–crank mechanism with fixed slider, used in a hand-pump.

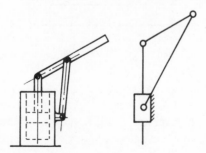

This process of changing the fixed links is called *inversion*. To distinguish it from the process of interchanging piston and cylinder it is sometimes called *chain inversion*, since it involves selecting as frame one of the chain of components that make up the mechanism; interchanging the pair of components that make up a joint, such as piston and cylinder, is called *pair inversion*. Chain inversion is of importance for several reasons, some of which will appear later. At this stage it can be pointed out that (as just shown) inversion enables us to discover several new mechanisms if we have one mechanism to start with. These mechanisms will have a great deal in common, so that one of the inversions may provide an improved alternative to the original mechanism, just as the oscillating-cylinder engine has certain advantages over the fixed cylinder type. All the inversions share the same line diagram and have the same relative motion of their links, so an understanding of one of them provides a key to the others. We can, for example, deduce the properties of epicyclic gear-trains from the behaviour of an ordinary gear-train by means of inversion.

1.3   Some terms and definitions

Many recent publications on mechanism are incomprehensible to the non specialist because of the special terminology that has come to be used in this subject. Some of these terms will now be defined.

Dynamics (or mechanics) is the science that deals with objects in motion and the forces required to initiate and maintain motion. Part of this study is purely geometrical – for example, the relation between θ and ϕ in Fig. 1.3(a) depends (apart from elastic deformations at high loads) only on the link-lengths. In a well-designed mechanism this relation is, to a close enough approximation, independent of forces, velocities and accelerations. This geometrical part of dynamics is called *kinematics* while the remainder, that deals with forces, is called *kinetics*. A gear-ratio, for example, is independent of speed or load and is a matter of kinematics; while the torque required to accelerate a gear-train is a matter of kinetics. Relative motion in a mechanism depends on the way the components are connected together as well as on certain dimensions such as link-lengths, so kinematics includes the study of different types of connections – pivots, slides, cam-action, etc. In fact most studies in mechanism are concerned with kinematics rather than kinetics. This applies particularly to the sort of complex, slow-moving machinery (such as packaging machines) where the theory of mechanism has its main use.

In kinematics we use the term *kinematic analysis* to mean the investigation of a given mechanism. Drawing a simplified diagram, as in Fig. 1.4, is an example of kinematic analysis, and velocity and acceleration diagrams are also included in this part of the subject. The term *synthesis* means devising, by a scientific process rather than trial-and-error, a mechanism to provide specified motion. Doing this in a single stage is called *direct synthesis* but more often the work falls into the following separate stages: *type synthesis* which means deciding what type of mechanism to use, such as a linkage or a gear-train; *number synthesis*, which means investigating the various mechanisms of this type, obtained, for example, by connecting the links in a different order or by inversion; and *dimensional synthesis*, which means finding key dimensions such as link-lengths. Type and number synthesis can be grouped together in a subject called *systematics*.

The parts of a mechanism, regardless of their shape or the type of connections between them, are called *links* or *bars*. When each point on every link moves in a plane, and all these planes are parallel, the mechanism is called *plane*, or *planar*. The mechanisms illustrated so far are plane, together with most other mechanisms, such as spur gears, or helical gears with parallel shafts. If a mechanism is not plane it is called a *space*, or *spatial*, mechanism; bevel and worm gears are examples.

The part of a link in contact with another link is called an *element* – for example, a gear-tooth or the bore of a cylinder. The connections between links are made by pairs of elements coming into contact, and these connections are therefore known as *kinematic pairs*. The type synthesis of mechanisms is carried out by arranging various combinations of links and pairs, in a way that can be compared with the synthesis of chemical compounds from the chemical elements. Considering only plane mechanisms, the kinematic pairs fall into two classes – *lower pairs* and *higher pairs*. Franz Reuleaux, the founder of modern kinematics, originally defined lower pairs as having surface contact (for example, a journal and plain bearing, or a piston and cylinder); and higher pairs as having line or point contact (for example, gear-teeth, or a cam and follower).

These definitions are still used by some writers, but what really matters is not the detailed physical construction of the joint but the kind of relative motion it allows between the two links. It is more useful to define lower pairing as allowing one degree of freedom, and higher pairing as allowing two. Some examples of these two kinds of pairing are shown in Fig. 1.9. On the left there is a pivoted

Fig. 1.9. Kinematic pairs.

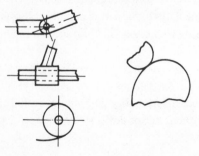

joint, a slider running on a guide, and a taut run of belt or chain leaving a pulley or sprocket. The taut run of belt is regarded as being able to pivot about its point of contact with the pulley, thus providing the same kind of connection as a pivoted joint. In these three cases the position of one link relative to the other can be described by a single dimension. For example, the relative position of the two links joined together by a pivot is determined by a single angle. Each of these three pairs therefore has a single degree of freedom and they are lower pairs. On the right-hand side of the figure is shown the type of contact found in a cam mechanism or in gearing. One link can slide and roll on the other and both these types of motion can be carried out either independently or together. Thus to define the motion of one link relative to the other we must specify both the motion of some point on that link and the angle through which the link has rotated. The pair has two degrees of freedom, and is a higher pair.

A *crank* is a link that rotates in a complete circle about a fixed pivot. A link that can only rock through a limited angle is called a *rocker* or *lever*. A link with no fixed pivot is called a *floating link* or *coupler* (there is one in the 4-bar linkage, and there are five in the linkage shown in Fig. 1.4).

A link having two elements, and hence being capable of connection to two other links, is called a *binary link*. One with three elements is called a *ternary link* and in general a link with k elements is called a *link of order k*. A pair connecting one link to another, such as an ordinary pivot, is called a *simple hinge*, one connecting a link to two others is called a *double hinge* and in general a pair connecting a link to k other links is called a *hinge of order k*. Links of order, 2, 3 and 4, and hinges of order 1 and 2, can be seen in Fig. 1.4.

The total number of links in a mechanism is denoted by n, and the total number of pairs by j. The number of links of order k is denoted by n_k, and the number of hinges of order k by j_k. Thus, for example, the mechanism in Fig. 1.4 has $n=10, n_4=1, n_3=3, n_2=6$; and $j=13, j_2=1, j_1=11$. Note that in counting the total number of pairs j a double hinge is counted twice, and in general a hinge of order k is counted k times. This does not apply to the links, each of which is counted once only. The reason for this distinction will appear later.

A set of links connected by kinematic pairs is called a *kinematic chain* (Fig. 1.10(a)). An ordinary chain is an example of a kinematic chain, using lower pairs if it is a roller-bush type driving chain and higher pairs in the case of an anchor chain. If every link is connected to at least two other links the chain is called a *kinematic closed chain* (Fig. 1.10(b)). Motion cannot be transmitted unless the chain is closed. If, in a closed chain, a particular link is chosen as a fixed link, i.e. as the frame, the chain becomes a *mechanism*. The significance of these important definitions can easily be understood by considering a 4-bar linkage. First of all, the links must be connected up to form a chain, and definite relations between the angles at the joints cannot be obtained until the chain is closed to form a quadrilateral. To make this closed chain into a usable device, one of the links must be fixed, say by holding it in the hand or clamping it in a vice.

We must now choose some definite way of driving the mechanism and once this is done it is known as a *motor-mechanism*. For example, the mechanism in Fig. 1.5 can be driven either by pressure on the piston as an engine, or by rotation of the crankshaft as a pump. Strictly speaking it is not a link that is driven, but always a pair of links — pressure on the piston requires an equal and opposite pressure on the cylinder and a torque on the crankshaft requires an equal and opposite torque reaction on the frame. This, of course, is simply an application of Newton's third law, that action and reaction are equal and opposite. The drive is thus applied, not to a link, but to a pair, called an *actuated pair*.

The *mobility* of a mechanism, denoted by M, is the number of inputs required to drive it. Most mechanisms (including all those shown in this chapter) have $M=1$; a differential gear is an example of a mechanism with $M=2$. It is not always easy to see how many inputs are required to drive a complex mechanism such as that shown in Fig. 1.4, but the value of M can be calculated from the *Chebyshev-Grübler formula*

$$M=3(n - 1) - j_h - 2j_l$$

where M = number of inputs, n = number of links, j_h = number of higher pairs, j_l = number of lower pairs. In this formula a hinge of order k must be counted k times. For example, the 4-bar linkage has $n=4$, $j_h=0$ and $j_l=4$, so $M=3(4-1)-2 \times 4 = 1$; and exactly the same calculation applies to the slider–crank mechanism. The Geneva mechanism (Fig. 1.6(c)), has $n=3$, $j_h=1$, and $j_l=2$, so $M=3(3-1)-1-2\times2=1$. The needle-bar drive (Fig. 1.4), has $n=10$, $j_h=0$ and $j_l=13$ (counting the double hinge twice), so $M=3(10-1)-2\times13=1$. There are exceptions to equation (1.1) but these, together with a proof and some further uses of this important formula, must be deferred to a later chapter.

A set of links can usually be assembled into a closed chain in several different ways. Each of these chains provides several mechanisms, obtained by different

Fig. 1.10. Kinematic chain and kinematic closed chain.

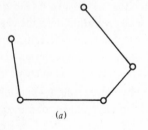

(a)

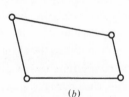

(b)

choices of fixed link. Each mechanism, in turn, provides several different motor-mechanisms. It will be shown later how the type synthesis of mechanisms can be carried out systematically on these lines, as a practical method of solving design problems.

1.4   Typical applications of linkages

Consider the three mechanisms shown in Fig. 1.11. The windscreen wiper, (*a*), provides an output motion in which the output link (the wiper blade) moves bodily – no point is fixed – and every point must move in a definite path. In the toy pedal-car drive, (*b*), the output is rotary. The output link is the driving wheel and one point on it, the centre, is fixed in the frame while every other point moves in a circle relative to the frame. Note the importance of defining the frame, relative to which motion takes place: relative to the ground the centre of the wheel moves in a straight line and every other point moves in a cycloidal path. In the cinema projector film-feed, (*c*), although the coupler has definite bodily motion, we really require only that one point, the claw engaging with the film, move on a specified locus. This is called point-path motion. The slider–crank mechanism (Fig. 1.5) has linear motion – the piston moves in a straight line.

Fig. 1.11. Some applications of the 4-bar linkage; (*a*) windscreen wiper (*b*) toy pedal-car and (*c*) cinema projector film-feed.

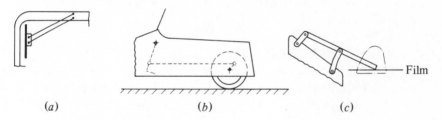

(*a*)   (*b*)   (*c*)

Fig. 1.12. Straw-baling machine using two slider–crank mechanisms.

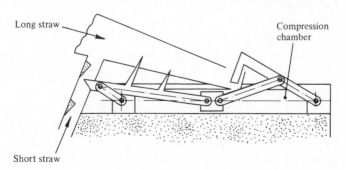

The three mechanisms shown in Fig. 1.11 are all 4-bar linkages, so it can be seen that one mechanism can give three different types of motion. The 4-bar can also provide a point-path which is a very close approximation to linear motion. Several outputs can be taken from a single mechanism at the same time, as in the straw-packer shown in Fig. 1.12. The left-hand slider–crank uses both rotary and point-path motion to feed straw into and along the chute; the right-hand slider–crank uses rotary motion to press the straw down into the compression chamber and linear motion to compress it into a bale. Not only can a single mechanism give several different types of motion but it can also be used for totally different purposes – the slider-crank, as we have seen, appears as an engine, a pump, a straw-packing machine and in countless other applications.

There are a number of handbooks that provide lists of mechanisms classified according to their uses. Helpful though these books often are, the examples given above show that attempts to classify mechanisms in this way cannot avoid overlap and confusion. The cataloguing and classification of mechanisms is best carried out by considering the way they are constructed. The selection of a type for some particular job is another matter altogether.

It is always useful in engineering to stop every now and again and ask ourselves 'What are we really trying to do?'. When designing a mechanism we are trying to achieve a particular sort of motion. The designer should begin by forming as clear an idea as possible of what type of motion he requires, say by sketching the output member in a sequence of positions, before he starts to consider the type of mechanism to be used.

1.5   Direct synthesis of mechanisms

A clear definition of the required output motion sometimes leads directly to a type of mechanism together with the main dimensions – the design is deduced logically from the specification without the intervention of intuition or experiment. This process is called direct synthesis.

Example. Fig.1.13(a) shows a simple device for guiding a component tray from a conveyor on the left to a chain-dotted position on the right, next to a machine tool. The tray is mounted on a link which swings about the pivot P. This pivot can be located by the construction shown in detail in Fig. 1.13(b). Mark two points A and B on the tray, and join them by a line. This line by itself is sufficient to show the position of the tray. Draw the line AB in the two required positions, denoted A_1B_1 and A_2B_2. Bisect A_1A_2 by a line a, and B_1B_2 by a line b. Then a and b intersect in the required pivot-centre P. It is an easy exercise to prove that triangles PA_1B_1 and PA_2B_2 are congruent, which shows that P, A and B can all lie in the same rigid link which will guide AB through the two specified positions.

We have, of course, shown quite generally that any plane motion between two specified positions can be carried out by rotation about a fixed centre unless *a* and *b* are parallel. In the latter case the motion is called a *translation* and can only be achieved by rectangular guides or a parallel action. The point *P* is called the *pole*. For reasons that will appear later we regard the possibility of *a* and *b* being parallel as a limiting case and speak of the pole as being at infinity.

It may happen that the pole lies in some inconvenient position so that the simple solution of Fig. 1.13(*a*) cannot be used. Note first of all that we cannot obtain a different pole by choosing a different line *AB*; the pole is unique, for if the body were pivoted at two distinct points it could not move at all. There is, however, an alternative solution. Take any point on *a* as fixed pivot for a link joined to *A*, and any point on *b* as fixed pivot for a link joined to *B*, to give the 4-bar linkage shown in Fig. 1.13(*c*). This will give the required guidance, together with considerable freedom of choice in practical design.

Fig. 1.13. How to locate the pole of a displacement, and use it for the synthesis of a simple mechanism.

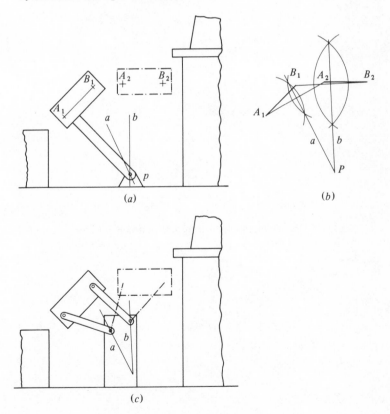

A link can be guided through three specified positions A_1B_1, A_2B_2 and A_3B_3, by the method shown in Fig. 1.14, which is an obvious extension of the two-position guidance just described. Bisect A_1A_2 and A_2A_3, and let the bisectors intersect in A_0. Then A_0 is the centre of a circle through A_1, A_2 and A_3, and can be used as fixed pivot for a link connected to A. Similarly, by bisecting B_1B_2 and B_2B_3, we find the other fixed pivot B_0 and construct a 4-bar linkage to give the specified sequence of positions. For specified moving pivot-centres A and B this construction gives only one solution, but by taking different positions for the moving pivots we can obtain different fixed pivots. A method of carrying out the design when the *fixed* pivots are specified will be given in Chapter 2.

Fig. 1.14. Design of a 4-bar linkage for three-position guidance.

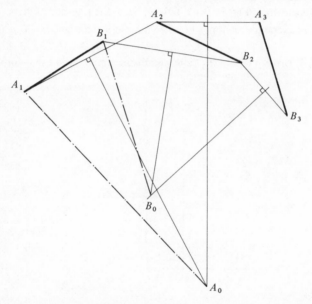

Fig. 1.15. Replacement of a long link (shown by chain-dotted lines) with a curved slot.

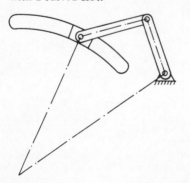

It may happen that the three positions of one of the moving pivots fall on a circle of large radius, or even on a straight line. In this event a 4-bar linkage will be impracticable or impossible. The difficulty can be overcome by guiding the pivot in question by a slot, as shown in Fig. 1.15, where the chain-dotted lines show the omitted links. Note that the sliding block in its slot is really a segment of a large pivot, and that when the slot is completely straight the 4-bar linkage becomes, in the limiting case, a slider–crank mechanism.

A slider in a guide is always more expensive and troublesome than a pivot, and in the case illustrated in Fig. 1.15 can usually be avoided by selecting another point on the moving link as a pivot-centre. This leads to the very useful idea of replacing a slider by a pivoted link even in a mechanism where a slider would normally be used. A good example is provided by the 'lazy tongs' bracket shown in Fig. 1.16 which is often used as an expanding towel-rail or adjustable telephone stand. Pick any point on the link which incorporates a slider and plot a sequence of positions, as in Fig. 1.16(*a*). Now draw a circle which is a reasonably good fit through the set of points, and find its centre. This locates the appropriate fixed pivot, giving the design shown in Fig 1.16(*b*). The success of the design depends on having the experience (and luck) to choose a point that lies on or near a circle of reasonable radius. In general, no point on the moving link in the original design will lie exactly on a circle throughout the motion, but the modified design will often give an output motion which is a close enough approximation.

Fig. 1.16. Replacement of a slide by a pivoted link.

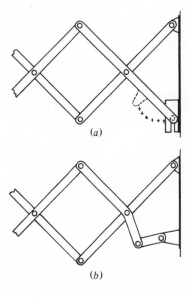

Exercises 1

1.1. Sketch conventional diagrams of the mechanisms shown in Fig. 1.17 and calculate the mobility of each of them.

Fig. 1.17.

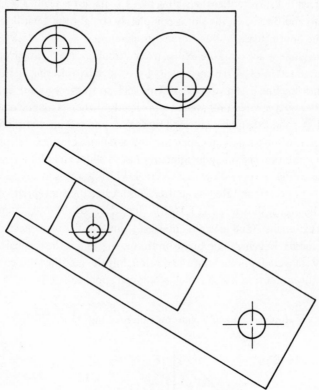

1.2. Sketch all the inversions of the kinematic chain shown in Fig. 1.18. Do you know of, or can you suggest, practical uses for any of them?

Fig. 1.18.

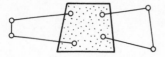

1.3. The link shown in Fig. 1.19 is to be guided from position 1 to position 2 by rotation about a single fixed pivot. Locate the pivot.

Fig. 1.19.

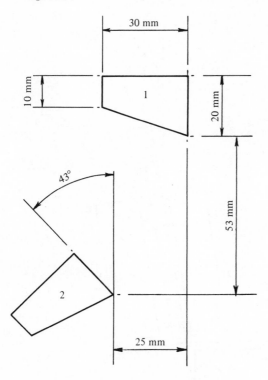

1.4. The back of a railway carriage seat is to be reversible, as shown in Fig. 1.20, so that the passenger can always sit with his back to the engine. Design a 4-bar linkage for this purpose, using the marked locations *A* and *B* for the moving pivot-centres. The fixed pivots may lie anywhere in the shaded area.

Fig. 1.20.

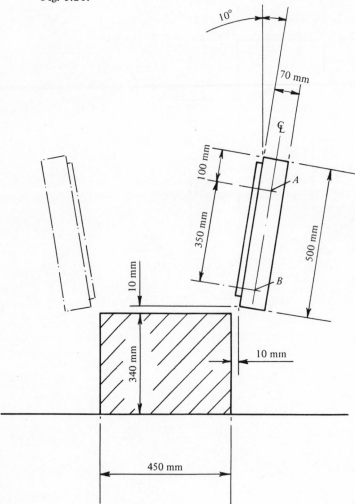

1.5. Design a lazy tongs bracket having pivoted joints only, as in Fig. 1.16(*b*). An extension of 600 mm is required. When closed the angle between adjacent links must not be less than 15° and the vertical distance between pivot-centre lines must not exceed 100 mm. When the bracket is fully extended the angle between adjacent links must not exceed 120°. When you have completed the design plot the path of the pivot-centre farthest from the fixed link.

1.6. Fig. 1.21. shows an up-and-over door in its open and closed positions; as the door moves between these positions it must not penetrate the shaded area. Design a 4-bar linkage to guide the door. (Hint: first make a design with the top of the door guided by slideways.)

Fig. 1.21.

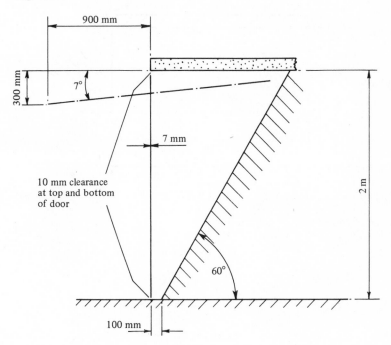

1.7. Fig. 1.22 shows a door in its closed position. Design a mechanism which will guide the door from this position to an open position allowing as much access to the cavity as possible. The mechanism must lie entirely inside the cavity and should be as small as possible so as to reduce obstruction. (Various mechanisms for this purpose are sold by ironmongers for use as cupboard door hinges; other examples can be found on car bonnets and boot lids).

Fig. 1.22.

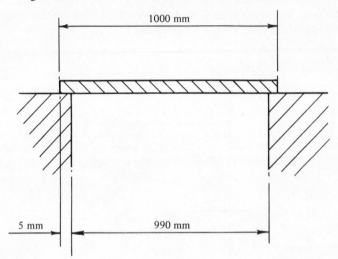

2 SOME ADDITIONAL DESIGN TECHNIQUES FOR THE 4-BAR LINKAGE

2.1 Use of given fixed pivot-centres

The design of a 4-bar to give specified bodily motion using specified pivot-centres on the moving link was described in Chapter 1. Practical considerations, however, may restrict the location of the fixed centres and make it necessary to use a different technique in which the fixed centres, instead of the moving, are specified at the start. The technique involves a principle which has other important applications as well. To fix ideas, consider the design of a linkage to enable a car bonnet to be lifted. To provide maximum access to the engine it is necessary to locate the fixed pivots as far back as possible.

The obvious approach is to treat the bonnet as fixed link and the car as moving link, enabling us to use the construction already described in Chapter 1. This is a case of inversion, i.e. of changing the fixed link, and we must now consider the practical problem of drawing the inverted sequence of relative positions of the two links, when the original sequence is our starting point.

Let us first emphasise a rather obvious notion taken for granted in Chapter 1. As shown in Fig. 2.1 (upper diagram) a single line with labelled end-points defines the position of a link, for the whole of the rest of the link can, if necessary,

Fig. 2.1. The position of a link can be defined by a single line with labelled end-points.

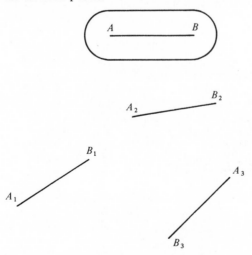

be located by pricking through points on a drawing of the link, this drawing being made on a small piece of tracing paper. Thus a sequence of positions of the link can be indicated, as in the lower diagram, by a sequence of lines. Note the importance of labelling the end-points of the line; in position 3 the link has been turned through nearly 180° relative to position 1. The defining line (often just called the link, since it stands for the actual component) is usually a line joining two bearing-centres, but of course it need not be. Any line in the link will do, for example one of the edges. Using this simplified representation of the links, the simplest technique for drawing an inverted sequence is as follows (Fig. 2.2).

Given a fixed link AB and two positions of a moving link CD (upper diagram), to invert the motion so that CD is fixed in position 1 and AB has the correct sequence of positions relative to it, mark $C_1 D_1$ as the new fixed link, and AB as the first position $A_1 B_1$ of the new moving link. Now trace the original positions of AB and $C_2 D_2$ on a small piece of tracing paper. Put this tracing over the drawing so that the trace of $C_2 D_2$ coincides with $C_1 D_1$, and prick the trace of AB through onto the drawing to give the inverted position $A_2 B_2$. We now have the inverted sequence drawn on the original sequence. For most applications it is best to have the inverted sequence on a separate sheet; it can easily be transferred by tracing, and in our example it is shown separately, on the lower diagram. We could deal similarly with a third position, and as many more as required; any sequence of relative positions can thus be inverted.

Figure 2.2. Inversion of a sequence of relative positions – an important first step in many design techniques. The upper diagram shows the sequence with AB fixed; the lower shows the same sequence of relative positions with CD fixed.

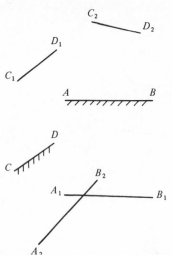

The idea, and actual technique, of inversion is very important and really quite simple. In position 1 we have left things as they were; for position 2 we have maintained the correct position of the two links relative to each other by means of our tracing. The reader should try it for himself and verify that it works by showing, with tracing paper, that the two links do in fact occupy the same position relative to each other when the sequence is inverted.

When the inversion has been performed we then proceed exactly as described in Chapter 1, treating AB as the moving link and locating on the paper pivot-centres whose positions in the actual moving link are defined in relation to the datum line CD. Another application of inversion will be given in the next section.

2.2 Function generation

Providing a specified rotary output from a given rotary input is known as function generation, because the technique was at first used mainly in the design of mechanical analogue computers in which it was necessary to make the output a specified mathematical function of the input. There are, of course, innumerable other uses for a rotary–rotary transmission. A good example is provided by the problem of controlling a valve on a machine using a handle located some distance from the valve spindle. A pair of spur gears would provide such a transmission, but if motion has to be transmitted over a distance they would be large and expensive. So also would gear-sectors. Bevel gears and a cross-shaft are often used for this purpose, but a much simpler and cheaper solution can be obtained with a linkage.

Suppose first that only two relative positions (pairs of input and output angle) are specified, as in the left-hand diagram of Fig. 2.3. Assume a length for the output link B_0B, and assume locations for the pivot-centres A_0 (input) and B_0 (output). Now invert the sequence with the input link fixed in position 1

Fig. 2.3. Use of inversion to design a linkage to connect two rotating shafts.

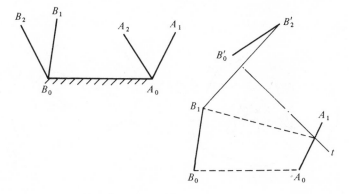

(right-hand diagram). The inverted position of the output link in position 2 is now $B_0'B_2'$. Join B_1 and B_2', and bisect with a line t. Then the unknown pivot-centre is located in link A_0A, anywhere on t. In the figure, where the completed design is shown by broken lines in the right-hand diagram, this pivot-centre is marked on the *line* A_0A_1, but it can be anywhere on the link so long as it lies on t. The procedure for three relative positions is an obvious extension. Note that since the shafts that rotate can be located however we please in their links, we are really arranging the transmission of one angular displacement rather than two angular positions.

Three-position synthesis, crude as it may seem, suffices for a vast number of practical designs. The result depends on choice of position, and a good approximation is usually obtained if we select one position in the middle of the range of motion to be transmitted, and the other two at about 8 per cent of the range from each end. Four, and sometimes even five, positions can be obtained but the design technique is considerably more advanced. There is, however, an approximate method of designing a function generator (called the overlay method) which can be used for multi-position synthesis and is remarkably easy. It is shown in Fig. 2.4.

By way of example we take a five-position synthesis (four specified angular displacements). Choose (arbitrarily) lengths for the input link and coupler. Draw the input link in the required sequence of positions and lay off circular arcs for

Fig. 2.4. The overlay method of 4-bar linkage design.

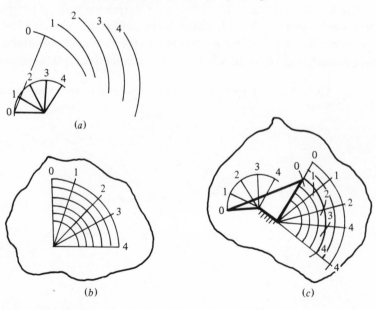

the far end of the coupler (Fig. 2.4(*a*)). Now take a small piece of tracing paper (called the overlay) and draw the required sequence of output link positions, together with a number of circular arcs marking possible lengths of this link (Fig. 2.4(*b*)). Put the overlay on the drawing and move it about until the arcs that represent the end of the coupler all cut the required output positions on the same circular arc. This completes the design, and the actual linkage has been drawn in heavy lines (Fig. 2.4(*c*)). The linkage can, of course, be altered in scale to fit specified fixed pivot-centres. Note that an exact fit is not obtained but, as in Fig. 2.4(*c*)), we can distribute the error as we choose. The method can be used for as many positions as required. If a satisfactory fit is not obtained a new trial should be made with a different coupler length. This technique, although simple and rapid, can be surprisingly accurate if the drawing is done carefully with a sharp pencil on film.

2.3   Design of crank–rocker linkages

The function generator is designed for limited rotation of both links. In a crank-rocker linkage one link rotates continuously while the other rocks to and fro. Such a mechanism has many uses, of which the most common is probably as a treadle drive, for example on a toy pedal-car or a sewing machine. It has been used in an electric shaver to swing the cutting head to and fro while the motor rotates continuously. The objective is to design a linkage in which the rocker moves through a specified angle ϕ while the crank moves through a specified angle θ. The rocker then returns to its original position while the crank rotates through the rest of its cycle, namely $360° - \theta$. Usually we wish to make the rocker advance slowly while it is doing work, and return quickly while it is running light, so it is necessary to have $\theta > 180°$. Thus the linkage in its two limit positions will appear as in Fig. 2.5(*a*). In these positions the crank and rocker are in line with each other, their lengths adding together in position 1, and the length of the

Fig. 2.5. Design of a crank–rocker linkage.

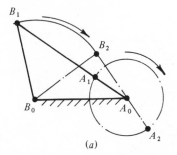

(*a*)

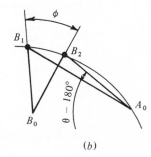

(*b*)

crank being subtracted from that of the rocker in position 2. Hence, with the notation in the figure,

$$A_0A + AB = A_0B_1$$
$$-A_0A + AB = A_0B_2$$

where AB is the length of the coupler, and A_0A the length of the crank. Solving for these two unknown link-lengths, we get

$$AB = \tfrac{1}{2}(A_0B_1 + A_0B_2)$$
$$A_0A = \tfrac{1}{2}(A_0B_1 - A_0B_2)$$

(2.1)

We are now in a position to design the linkage, using the construction shown in Fig. 2.5(b). Choose a length for the rocker and a centre B_0, and draw the rocker in the required limit positions. Now draw a pair of lines through B_1 and B_2 to intersect at an angle $(\theta - 180°)$. The point of intersection is A_0, and the actual lengths of crank and coupler can be obtained from equations (2.1).

This completes the design, but we can usefully take the matter a little further. Draw a circle through A_0, B_1 and B_2. Then, by a well-known geometrical theorem, any point on the circumference of this circle will serve as an alternative location for A_0. We thus have a whole family of alternative designs. We must choose between them, and that is the subject of the next section.

2.4 Transmission angle

The transmission angle is defined as the acute angle between the coupler and output link of a linkage (Fig. 2.6(a)). In a slow-moving mechanism, where the forces can be calculated by statics, the maximum torque on the output link for a given force acting along the coupler occurs when this angle is 90°, and no torque at all will be obtained when the angle is zero. In practice the linkage is likely to be troublesome if the transmission angle falls much below 45°. Thus the design must

Fig. 2.6. Transmission angle; it is the acute angle between the coupler and the output link. In a crank–rocker mechanism the minimum value of transmission angle is found at one of the two positions when the crank is in line with the fixed centres; both must be tested to ensure that this angle is not too small.

Transmission angle

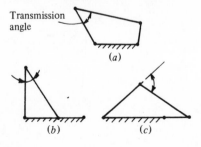

(a)

(b)   (c)

be checked for the minimum value of transmission angle. For double-rocker linkages it is best to draw the linkage in a few positions and measure the angle with a protractor. In the case of a crank–rocker the minimum angle will occur in either of the two positions in which the crank is in line with the fixed pivot-centres (Fig. 2.6(*b*) and (*c*)). Draw both, and pick the worst case. If the transmission angle is too low, try another position for A_0 on the circle described in the last section.

Practical use can be made of a small or zero transmission angle to obtain a locking action, as in the clamp shown in Fig. 2.7 – an excellent example of clever use of a 4-bar linkage. It can be seen that when the clamping bar is horizontal the coupler lies along the handle. Regarding the handle as output link working against a load supplied by its own pivot friction, it is clear that no force applied to the clamping bar can open the clamp.

Fig. 2.7. Toggling action of a clamp – a useful application of the fact that a mechanism will jam if the transmission angle is very small.

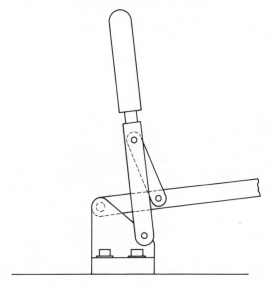

Exercises 2
2.1. Fig. 2.8 shows the open and closed positions of the guard on a machine tool. As the guard is moved between these positions it must not penetrate the shaded area, otherwise it might strike the operator. The only available positions for fixed pivot-centres on the machine are those marked *A* and *B* on the figure. Design a 4-bar linkage to control the motion of the guard.

Fig. 2.8.

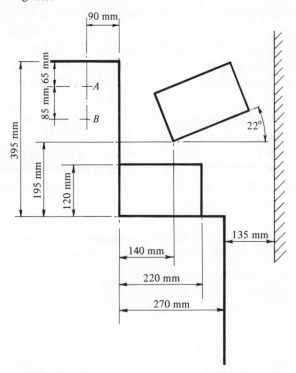

2.2. Two shafts are 100 mm apart on a horizontal line. The left-hand shaft is to rotate through 60° clockwise while the right-hand shaft rotates through 90° counter-clockwise. Design a 4-bar linkage to connect them. Draw the linkage in a few intermediate positions. Taking the right-hand shaft as input member plot a graph of transmission angle against input angle. Estimate the worst transmission angle from your graph, and modify the design to improve it if it is below 45°. Finally, plot a graph of output angle against input angle and estimate the maximum error as compared with spur gearing.

2.3. The windscreen wiper on a car is required to swing through 120°, being driven by a continuously rotating link of length 30 mm. The centre-distance between the wiper and driving link is 100 mm and they are to be connected by a coupling rod so as to form a crank–rocker linkage. Find the distance between the centres on the wiper arm, and the minimum transmission angle.

3 POINT–PATH MOTION, THE SLIDER–CRANK MECHANISM AND THE 6-BAR LINKAGE

3.1   Designing a 4-bar linkage for point-path motion

A great variety of curves can be traced by points carried on the coupler of a
4-bar linkage. At each end of the coupler, i.e. at the pivot-centres, the path is of
course a circular arc. Elsewhere it is defined by a lengthy algebraic equation of
the sixth degree. Formal synthesis of a linkage to give specified point-path motion
is more difficult than the other types of synthesis described in this book. A good
approximate method will now be described.

Suppose (to take a definite example) we wish to design a film-transport of the
type shown in Chapter 1 (Fig. 1.11). Choose a location for the crank pivot (driving
link) and sketch the required path, and the coupler with its claw, as in Fig. 3.1(*a*)
Now trace the coupler on a piece of tracing paper with a fair amount of sur-
rounding area. By placing this over the drawing (Fig. 3.1(*b*)) we can prick through
and trace on the drawing the locus of any coupler point as the mechanism runs
through its cycle. Try a point that looks as though it may move approximately
on a circular arc, and trace the locus. If the locus is in fact a fair approximation
to a circular arc, and runs through about 90° of a circle, the point can be used as
the remaining coupler pivot-centre, and the centre of the arc as the remaining

Fig. 3.1. Design of a 4-bar linkage to give specified point-path motion.

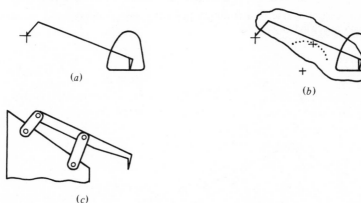

(*a*)

(*b*)

(*c*)

fixed centre (Fig. 3.1(*c*)). The point may not lie within the area of the original sketch of the coupler, which is why the cut-out should have a surround. The locus will be only an approximation to a circular arc, so the performance of the resulting linkage should be checked by plotting claw positions throughout the cycle – or, better still, by means of a cardboard model using drawing pins as pivots. Since we can usually allow considerable latitude in the shape of the point-path the simple method described here is often successful. It is better than more complex methods in one very important practical way – we can see, in the course of the procedure, just how the whole mechanism will behave.

3.2 Dimensional synthesis of slider–crank mechanisms

The simplest case is that of a reciprocating engine or pump in which the crank-centre lies on the cylinder axis. The length of the crank must be half the required stroke. The connecting-rod should be as long as possible to reduce cylinder wall loading, but its length is limited in practice by considerations of inertia and overall size.

Another case, almost as simple, is mentioned here because some designers who meet it for the first time are puzzled by it. This is the use of a trunnion-mounted hydraulic or pneumatic cylinder to swing a link through a specified angle for a specified stroke (Fig. 3.2). Draw the link in the required two positions. Bisect the angle between them by a line *t* and draw *PR* of length equal to half the stroke, perpendicular to *t*. Now draw a line through *R* perpendicular to *PR*. This line will cut the link at the correct pivot-centre for attaching the piston-rod. The cylinder axis must be parallel to *PR* and the trunnion can lie anywhere on that axis.

Three-position synthesis of a slider–crank can be done by inversion (Fig. 3.3). Mark required corresponding positions of crank and slider (Fig. 3.3(*a*)). Now

Fig. 3.2. Use of a trunnion-mounted ram to swing a link through a specified angle.

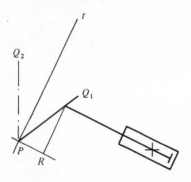

invert, with the crank fixed (Fig. 3.3(*b*)). Finally, locate the centre of the circle through the three new positions of the slider, as in Fig. 3.3(*c*), where the actual mechanism is shown superposed in broken lines.

An overlay method can also be used, as in Fig. 3.4. Four positions (for example) of the cross-head are marked off, together with circular arcs to indicate the endpoint of a connecting-rod of some assumed arbitrary length (Fig. 3.4(*a*)). An overlay for the crank is prepared on a small piece of tracing paper (Fig. 3.4(*b*))

Fig. 3.3. Design of a slider–crank linkage by inversion.

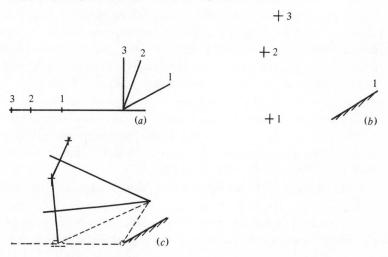

Fig. 3.4. Design of a slider–crank linkage by the overlay method.

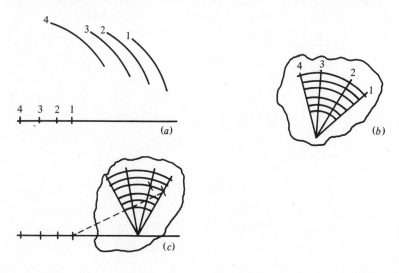

and superposed to obtain a good fit (Fig. 3.4(*c*)), just as in the overlay method
for the 4-bar linkage in Chapter 2. In this case the line of travel of the cross-head
centre will not necessarily pass through the crank-centre, whereas in the inversion
method we can always ensure that it does.

The synthesis of a slider–crank can be an intermediate step in a design which
will end up using pivoted links only. The following problem provides an example.
The movement of a certain type of voltmeter gives a rotary output which is not
proportional to the applied voltage, but some non-linear function of it, so that
the scale would be cramped at one end. This effect must be compensated for,
and the output displayed by a needle moving in a straight line instead of round
a circular scale. Corresponding positions of the rotary instrument action and the
needle position are shown in Fig. 3.5(*a*). The first step is to synthesise, say by the
overlay method, a slider–crank to correlate these positions (Fig. 3.5(*b*)). But we
cannot use this design as it stands. Slider friction cannot be allowed in an instrument, and even if it could this design has very poor transmission angles and would
undoubtedly jam. We overcome the difficulty by the method already used in
another example (Chapter 1, Fig. 1.16). Find a point on the connecting-rod that
moves on a fair approximation to a circular arc, find the centre of that arc, and
transform the design into a 4-bar linkage (Fig. 3.5(*c*)).

It may seem that in the last example simplicity has been taken too far — a
rough graphical construction adequate for the design of an expanding towel-rail
is surely not good enough for a precise instrument. In fact the accuracy of results
achieved by these methods is often surprisingly good, and may be more than

Fig. 3.5. Design of an instrument mechanism to give a linear output on
a straight scale, from a non-linear rotary input. First design a slider-
crank linkage and then convert it to a 4-bar linkage.

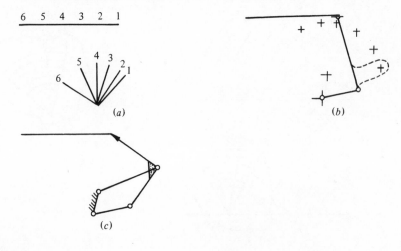

(*a*)

(*b*)

(*c*)

enough for an instrument such as a car thermometer or oil pressure gauge which is only required to give a rough indication. It is, in any case, useful to be able to produce quickly an approximate design so that we can investigate feasibility, cost and overall size before undertaking the labour of producing a more exact solution.

3.3 Introduction to 6-bar linkage design

The two basic types of 6-bar linkage are shown in Fig. 3.6. It will be shown in a later chapter that a large number of different mechanisms, with many applications, can be derived from them, mainly by inversion. Meanwhile it should be noted that the type shown in Fig 3.6(*a*), known as Watt's linkage, is simply one 4-bar driving another, while the type shown in Fig. 3.6(*b*) (Stephenson's linkage) is something essentially new.

Possibly the most interesting and important use of these linkages is to obtain an oscillatory output with a very close approximation to a dwell. Motion of this type, which is often required in textile machinery, was at one time usually obtained by a cam, but the greatly increased speeds required in modern machines cannot be achieved by cam mechanisms. This is mainly because of the high load

Fig. 3.6. The two basic forms of the 6-bar linkage – Watt's linkage (*a*) and Stephenson's linkage (*b*).

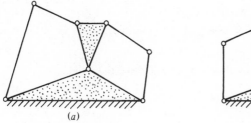

(*a*)   (*b*)

Fig. 3.7. An 8-bar linkage giving rotary motion with a dwell. This old design can be replaced by the simpler 6-bar types shown in Fig. 3.8.

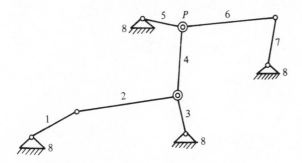

and rubbing velocity which occur at the line-contact between cam and follower, which would give rise to rapid wear.

Early attempts to use a linkage as a dwell mechanism were based on a train of 4-bars driving one another, usually resulting in an 8-bar linkage as shown in Fig. 3.7. This linkage is arranged so that link 1 is the driving link of a crank–rocker with link 3 as rocker. At the extreme right-hand position of the rocker, when the rocker is moving very slowly, links 3 and 4 come into line and the pivot *P* is almost stationary. At the same time links 5 and 6 come into line, so that the output link 7 practically ceases to have rotary motion. In such a design the output link can be practically stationary for as much as a third of the cycle. It is easy to understand the action of this mechanism if one makes a simple cardboard model.

Dwell mechanisms using Stephenson's linkage offer considerable advantages over the 8-bar just described. They use six links instead of eight, fewer bearings, and can be designed to give a closer approximation to an exact dwell, two dwells per cycle if required, and minor refinements such as a dip during a dwell. There are two ways of designing such a mechanism.

Suppose first that the pivot *P* in the Stephenson's mechanism shown in Fig. 3.8(*a*) moves on the path shown, which has an approximately circular portion. Let the link *PQ* have a length equal to the radius of curvature of this part of the coupler curve. Then the point *Q* will be practically stationary, and the output link *QR* will have an approximate dwell. In practice the dwell can be very good; in a model I tested it was necessary to use a dial gauge to detect motion during

Fig. 3.8. Rotary motion with a dwell using 6-bar linkages.

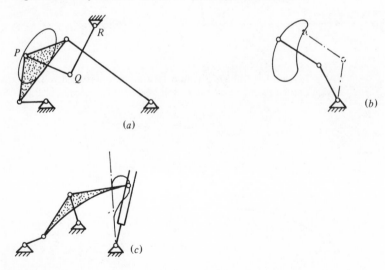

(a)

(b)

(c)

an approximate dwell lasting for 130° of the cycle. It is also possible to arrange for a coupler point to have the sort of path shown in Fig. 3.8(*b*), which has two portions with approximately circular arcs of the same radius of curvature. Such a mechanism will give two dwells.

A similar mechanism using a slider and only five links is shown in Fig. 3.8(*c*). Here we arrange for the coupler curve to have an approximately straight portion, and mount the slotted output link so that the axis of the slot coincides with this part of the coupler curve. It is easy to see that in the position shown by chain-dotted lines in the figure the output link will have an approximate dwell.

EXERCISES 3

3.1. A link *AB* (Fig. 3.9) is pivoted to the frame at *A* and to the piston-rod of an hydraulic cylinder at *B*. The other end of the cylinder is pivoted to the frame at *C*. The cylinder has a stroke of 60 mm and, when it is fully closed, *BC* = 180 mm. The link *AB* is required to swing through $20°$ and, when the cylinder is fully extended, *AB* is to be perpendicular to *BC*. Find the length of the link *AB* and the distance between the fixed centres *A* and *C*.

Fig. 3.9.

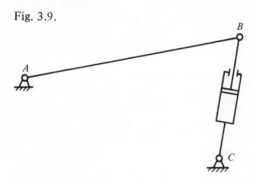

3.2. The machine guard controlled by a 4-bar linkage in Exercise 2.1 is to be opened and closed automatically by a pneumatic cylinder. Specify the stroke and pivot locations of a suitable trunnion-mounted cylinder, bearing in mind: (*a*) the restriction on fixed centres in the original question and (*b*) that the centre-distance on the fully closed cylinder is 200 mm.

3.3. Boxes 250 mm long are to be pushed along a track with 300 mm gaps between them by claws passing through slots in the track, as indicated in Fig. 3.10. Design a linkage to drive the claws.

Fig. 3.10.

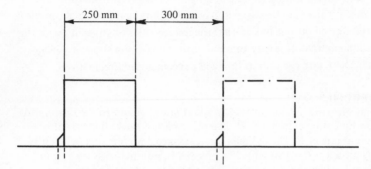

3.4. For a mechanical recording instrument it is necessary to have a mechanism that converts the rotation of a shaft into the approximately linear motion of a pen, with displacements of the pen approximately proportional to angles of rotation of the shaft. The ideal result is one in which deviations from linearity and proportionality are less than the eye would detect.

Achieve the best solution that you can with a 4-bar linkage, as indicated in Fig. 3.11, by which 90° of clockwise rotation of the right-hand crank gives 110 mm of horizontal motion, from left to right, of the coupler point Q. The fixed centre of the right-hand crank should be approximately 100 mm below the right-hand end of the line of travel of the point Q.

Fig. 3.11.

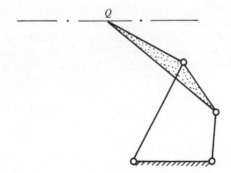

3.5. Design a 6-bar linkage to give an output oscillation of $45°$, with a dwell at one extreme of the oscillation during $80°$ of rotation of the driving crank. The crank must be capable of continuous rotation, i.e. the 4-bar group must be of the crank—rocker or double-crank type. No transmission angle must fall below $40°$. Plot graphs of the output, and of the two transmission angles, as functions of the driving crank angle.

4 POLES, POLODES, VELOCITY DIAGRAMS AND CONTOUR CAMS

We have now considered the simplest mechanisms and the simplest techniques for designing them. The purpose of a mechanism is to make things move in some specified way. It is therefore useful, before considering more complex mechanisms and design techniques, to look a little more closely at motion itself.

We have so far treated motion as a sequence of positions, usually three. In this chapter we consider continuous motion and velocity.

4.1 Poles and polodes

It was shown in Chapter 1 (Fig. 1.13) that any plane motion from one position to another can be achieved by rotation about a fixed point called the *pole*. It is convenient to refer to the positions by number, 1 and 2, and to the pole as having position P_{12} in the fixed link and position P'_{12} in the moving link. Now suppose there are further movements to positions 3, 4, etc. Each of these can be achieved by rotation about poles having positions P_{23}, P_{34}, etc. in the fixed link or frame (designated P'_{23}, P'_{34}, etc. with respect to the moving link). This is shown in Fig. 4.1(a). Now suppose we draw all the P' positions on the same view of the moving link (by triangulation or the use of tracing paper) as in Fig. 4.1(b). Finally, if we join up the pole positions in each link by straight lines (called *pole-polygons*) as in Fig. 4.1(c), we obtain the outlines of a pair of links which would have the specified relative motion if one were rotated about the other at each pole in succession. Note that, as usual, we are concerned only with relative motion between the two links, and it is purely a matter of convenience which we call fixed and which moving. In practice both could move, and we could construct the poles after an inversion to fix either in any one position. It would be possible to construct actual links, as in Fig. 4.2, to give the specified motion but impact as each pole engaged would make such a device impracticable.

Suppose, however, that we take a greater number of intermediate positions, and eventually an infinite number, so that the pole-polygons become smooth curves as in Fig. 4.3. In the case of pole-polygons the distance between two successive poles P_n and P_{n+1} is always equal to the distance between the corresponding points P'_n and P'_{n+1}. This equality is true no matter how many poles we take,

Fig. 4.1. Construction of poles and pole-polygons for a sequence of positions.

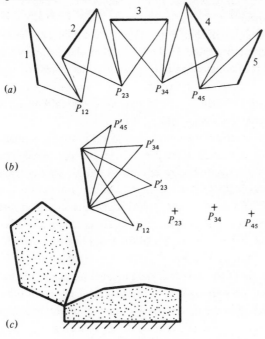

(a)

(b)

(c)

Fig. 4.2. Specified motion of a link could be achieved by rotations about successive poles.

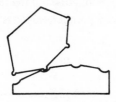

Fig. 4.3. Polodes – the limiting case of pole-polygons when we consider an infinite number of small displacements. The actual motion can now be achieved smoothly, by pure rolling.

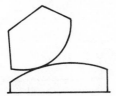

and it remains true in the limit when the pole-polygons become smooth curves. Hence the motion is one of pure rolling, without slip. These rolling curves are called *polodes*, or *centrodes*.

If we regard one link as fixed, then in the case of pole-polygons the point of contact P' on the other link is at rest. In the limit, when the pole-polygons become polodes, the point of contact has zero velocity at the instant of contact and at that instant the moving link behaves as though it were pivoted at the point of contact, which is therefore called an *instantaneous centre*; although modern writers tend to call it simply a pole, as indeed it is.

Polodes can be used directly, for example as a wheel in rolling contact with the ground or with another wheel, or as a geometrical concept to help our understanding of motion. To investigate the instantaneous motion we can replace the polodes (Fig. 4.4(a)) by circles with the same radii of curvature at the point of contact (Fig. 4.4(b)). The point P' in Fig. 4.4(a) is instantaneously at rest (i.e. it has zero velocity), but it does not have zero acceleration. It remains, of course, fixed in the moving link, but moves relative to the fixed link on the path shown in Fig. 4.4(b), which is initially a cycloidal curve. The initial motion of P' is along the common normal of the polodes. The moving link as a whole moves instantaneously as though it were pivoted at the pole, and any point other than P', for example the point Q in the figure, has a velocity of direction perpendicular to $P'Q$, in the sense of the rotation, and of magnitude proportional to the distance between P' and Q.

Suppose we draw the velocity of Q, which we denote as $\dot{Q}$, to any scale, and draw a line from P through the head of this vector, as in the figure. We represent the angle between $P'Q$ and this line by the special symbol ϑ used in the figure, which is a form of the Greek letter theta and is usually pronounced 'tetta'. Note

Fig. 4.4. Instantaneous centre, tetta-angles and velocities.

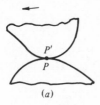

(a)

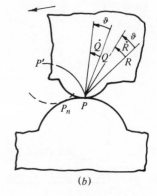

(b)

that this angle has the sense of the rotation. Then, because velocity is proportional to distance from P', we will have the same angle tetta for any point in the moving link, provided we draw all velocities to the same scale. So if we know two things about the motion of a link – the pole, and the velocity of one point – we can lay off the velocity of any other point R by the simple and rapid construction shown in the figure.

The moving link will eventually reach another position, as indicated by a broken line in the figure, where the pole in the fixed link will have some new position P_n. A small time is required to achieve this motion, and the distance PP_n divided by that time is an average velocity. In the limit, when the time tends to zero, we call this the *pole velocity*; it is, for example, the rate at which a wheel would mark the road if we chalked the rim.

The location of the (instantaneous) pole is often very simple. Recalling Fig. 1.13(b), the pole must lie on the perpendicular bisector of the line joining two successive positions of any point in the moving link. In the limit, as the two successive positions become closer and closer, the bisector becomes the normal to

Fig. 4.5. Velocity analysis of a 4-bar linkage by the use of the instantaneous centre (a); location of the instantaneous centre for a slider-crank linkage (b); and a 4-bar linkage in the crossed configuration (c).

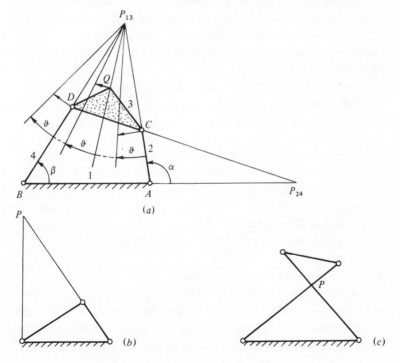

the path of that point. Consider the application of this fact to the 4-bar linkage (Fig. 4.5(a)). We know the paths of the two points C and D on the coupler – they both move in circles, about the fixed pivots A and B respectively. The normal to a circle is the radius, so the pole must lie on both AC and BD; hence it lies at their intersection P_{13}. Note that this notation, in the case of an instantaneous motion, means the pole of links 1 and 3, and not, as in the case of finite motion, the pole of a rotation between positions 1 and 3. Now suppose we know the angular velocity of a crank, say link 2. We lay off the velocity of C to some scale. We can then draw tetta-lines, and hence find the velocity of D (and from this the angular velocity of the other crank), and the velocity of any coupler-point, such as Q. The whole coupler behaves instantaneously as though it were a link pivoted at P_{13}. Note that if we wish to find the ratio of the *angular* velocities we do not need to specify the scale – it is required only if we wish to find a *linear* velocity, such as the velocity of Q, when we are given the angular velocity of one of the cranks.

The pole for the relative motion of links 2 and 4, namely P_{24}, is found at the intersection of BA with DC. The two cranks behave instantaneously as though they were spur gears having internal contact at P_{24}. Hence their velocity-ratio is given by

$$\iota_{24} = \dot{\alpha}/\dot{\beta} = BP_{24}/AP_{24}$$

which is a very simple, quick way of finding the angular velocities.

The case of a slider–crank mechanism is shown in Fig. 4.5(b), and is similar except that the normal to the connecting-rod at the cross-head is perpendicular to the axis of the cross-head guides.

Note that if a 4-bar is in a crossed configuration (as in Fig. 4.5(c)) the pole for relative motion of the two non-intersecting links lies at the intersection of the other two. It is also most important to remember that none of these diagrams is changed by inversion; we are concerned here with the motion of the links relative to one another, and it is of no significance which link is fixed.

It is becoming increasingly usual nowadays to denote a point by, for example, A, its velocity by $\dot{A}$, and its acceleration by $\ddot{A}$. This is much better than the old convention of one letter for position, another for velocity and a third for acceleration; there are fewer letters to remember. Similarly we denote an angle by, say, α, the angular velocity by $\dot{\alpha}$, and the angular acceleration by $\ddot{\alpha}$.

4.2 The Aronhold–Kennedy theorem
For any three links 1, 2 and 3 in instantaneous motion the three poles P_{12}, P_{23} and P_{31} lie on a straight line. This can easily be understood by examining Fig. 4.6. We regard link 1 as fixed and consider velocities relative to it. Now point P_{23} is really two coincident points, one on link 2 and the other on link 3. The

velocity of the first must be perpendicular to $P_{12}P_{23}$, and the velocity of the second must be perpendicular to $P_{31}P_{23}$. But these two velocities must be equal, since there is no relative motion between links 2 and 3 at P_{23}. Hence $P_{12}P_{23}$ and $P_{31}P_{23}$ must have the same direction, and the three poles must therefore lie on the same straight line.

The theorem is useful for finding the poles when there are floating links none of whose points have obvious paths, as in the case of the 6-bar linkage shown in Fig. 4.7(a). It is usual to keep a tally of the poles as they are found, by marking as many points as there are links, on a separate diagram (Fig. 4.7(b)). As poles are found the corresponding points on the diagram are joined up, until every point is joined to every other, which means all poles have been located. Permanent centres, i.e. pivots, are also poles. Seven of these can be marked off immediately on the linkage and on the auxiliary diagram, thick lines being used in the latter case. We can also easily locate P_{13} and P_{26} since they are poles of a 4-bar chain. They are indicated by thin lines on the auxiliary diagram. We now start to use the theorem.

Whenever we find a closed quadrilateral on the auxiliary diagram we can use the theorem to put in the diagonals. For example, points 1, 3, 4 and 5 are already

Fig. 4.6. The Aronhold–Kennedy theorem; if three links are in relative motion their three instantaneous centres must lie in the same straight line.

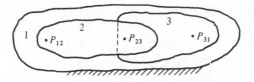

Fig. 4.7. Locating the poles of a 6-bar linkage.

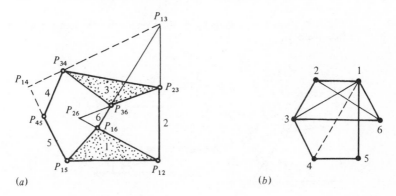

(a)   (b)

connected. Now P_{14} must lie on $P_{15}P_{54}$, and on $P_{13}P_{34}$. Hence it must lie on the intersection of these two lines, shown as broken lines on the linkage. Similarly we can find P_{35}. Now the four points 1, 5, 3 and 6 on the auxiliary diagram become joined by a crossed quadrilateral, of which the line joining points 5 and 6 is a 'diagonal' in the sense that obviously applies here. Hence we can find P_{56}. All the other remaining poles can similarly be found. They have been left out of the figure, which would otherwise be rather a maze of lines.

4.3   Practical solution of velocity problems

Poles cannot always be used; they may lie too far off, or be formed by lines that intersect at very acute angles and hence define them poorly. In such a case we can use ordinary velocity diagrams (discussed below), or a combination of both methods. An example is shown in Fig. 4.8. The figure shows an engine mechanism in which the connecting-rod is pivoted to a point on the coupler of the 4-bar linkage. We are given the piston velocity when the mechanism is in the position shown, and asked to find the angular velocity of the crank, link 2. An attempt to use the ordinary velocity-diagram method, starting with the piston velocity, will fail because we have two floating links (5 and 3) in succession; so will an attempt to use poles, because we only know the path of one point on link 5. The easiest way to solve the problem is to start from the other end, and draw a line to represent the velocity of point Q on link 2 to an unknown scale. As shown in the figure we find the pole P and use tetta-angles to lay off the velocity of point R. This gives us the direction of point R, and point T (the cross-head pivot) moves along link 1, which is fixed. So we can now locate the pole of link 5 relative to the frame and hence lay off the velocity of point T. But in the example this pole would be off the paper so we draw a small auxiliary velocity diagram as shown in the figure. We now have the velocity of point T, i.e. of the piston. Choose a scale to make this represent the given velocity; then we know the velocity of point Q, and hence the angular velocity of link 2, as required. We

Fig. 4.8. Velocity analysis by combining a velocity diagram with the use of poles.

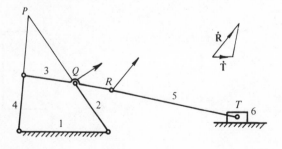

could work the whole problem by a velocity diagram, again starting from link 2, but the method described is more direct.

4.4 Velocity diagrams

The ordinary velocity-diagram method is well known, but not always well understood. It is worth looking at from first principles. To begin with we will establish the principle on which it is based, namely that the *difference* between the velocities of any two points on a rigid link is a vector perpendicular to the line joining them.

Note that we do not speak of the velocity of one point *relative* to another. Velocity has direction, and direction can only be measured relative to a line, as shown in Fig. 4.9.

To prove the basic principle we use complex numbers, with the notation shown in Fig. 4.10. We consider a link with two marked points, P and Q, with a distance a between them. The position vectors $\mathbf{P}$ and $\mathbf{Q}$ are represented by complex numbers, while a is a constant scalar. Then the position of Q is given by

$$\mathbf{Q} = \mathbf{P} + a e^{i\theta}$$

where θ is the angle that the line PQ makes with the frame, measured counterclockwise. To find velocities we simply differentiate with respect to time, obtaining

$$\dot{\mathbf{Q}} = \dot{\mathbf{P}} + i\dot{\theta} a e^{i\theta}$$

Hence

$$\dot{\mathbf{Q}} - \dot{\mathbf{P}} = i\dot{\theta} a e^{i\theta} \tag{4.1}$$

Fig. 4.10. A rigid link in plane motion.

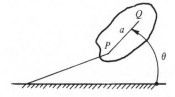

Fig. 4.9. Velocity must be measured relative to a line, not a point, because it is a vector.

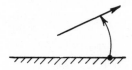

Thus the difference between the velocities of the two points is a vector perpendicular to the line joining them, of magnitude equal to the length of that line
multiplied by the angular velocity of the link.

It follows immediately that if a link rotates about a fixed pivot the velocity
of any point Q on the link has magnitude $a\dot{\theta}$ and direction perpendicular to the
line joining Q to the centre of rotation, where a is the distance from the centre
and $\dot{\theta}$ the angular velocity.

Let us now, in order to illustrate the principle, consider the simple example
of a velocity diagram for a 4-bar linkage (Fig. 4.11). Draw a line to represent
the velocity of P, mark it $\dot{\mathbf{P}}$, and put an arrow on it to indicate direction. Now
draw a line t of indefinite length through the head of this vector, perpendicular
to PQ. Finally, draw a line in the direction of $\dot{\mathbf{Q}}$ to meet t, mark it $\dot{\mathbf{Q}}$, and put an
arrow on it. All this is well known. There is, however, much confusion about
the line t, whose length is now determined and represented by the thicker part
shown in the diagram. It is a vector; but is it $\dot{\mathbf{P}}\text{-}\dot{\mathbf{Q}}$, or $\dot{\mathbf{Q}}\text{-}\dot{\mathbf{P}}$, and can we put an
arrow on it? The answer is simple. It is the third term in a vector sum (vector
triangle), and we mark it in either of the two ways shown in the two lower diagrams. We have done this without talking about the velocity of one point relative
to another, and without writing a vector equation. Such an equation is usually
unnecessary, but if we wanted to write it down it would take either of the forms

$$\dot{\mathbf{P}} + (\dot{\mathbf{Q}} - \dot{\mathbf{P}}) = \dot{\mathbf{Q}}$$

or

$$\dot{\mathbf{P}} - (\dot{\mathbf{P}} - \dot{\mathbf{Q}}) = \dot{\mathbf{Q}}$$

depending on which of the two alternative labellings in the lower diagrams we
preferred to use.

Fig. 4.11. Velocity diagram for a 4-bar linkage. The origin of co-ordinates
is at O.

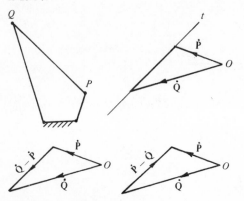

A slight addition to the theory is needed in order to treat the case of a link with a slider (Fig. 4.12). We consider the difference between the velocities of two coincident points, Q on the link and Q' on the slider. They are both distant x from a fixed point P on the link, but we must treat x as constant for Q and variable for Q'. Then

$$\mathbf{Q} = \mathbf{Q}' = \mathbf{P} + xe^{i\theta}$$

but

$$\dot{\mathbf{Q}} = \dot{\mathbf{P}} + i\dot{\theta}xe^{i\theta}$$

while

$$\dot{\mathbf{Q}}' = \dot{\mathbf{P}} + i\dot{\theta}xe^{i\theta} + \dot{x}e^{i\theta}$$

so

$$\dot{\mathbf{Q}}' - \dot{\mathbf{Q}} = \dot{x}e^{i\theta} \tag{4.2}$$

which means that the difference between the velocities is a vector of magnitude $\dot{x}$ along the link. This, of course, is what common sense would lead us to expect. But we cannot always rely on common sense in these matters, because the difference between the accelerations is not, as we would expect, $\ddot{x}$. Incidentally, we can

Fig. 4.12. The difference between the velocities of coincident points Q and Q' is in the direction of sliding; this is not true of accelerations, where the Coriolis component must be taken into account.

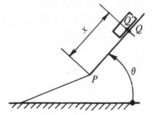

Fig. 4.13. Velocity diagram for a slider–crank linkage of the rotary engine type.

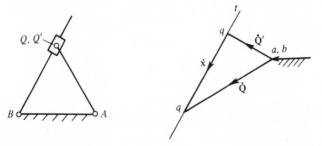

obtain the acceleration difference, including the Coriolis component, by differentiating $\dot{Q}$ and $\dot{Q}'$ and forming the difference $\ddot{Q}'\text{-}\ddot{Q}$; this is much easier and more reliable than the older textbook methods, and the signs are automatically correct.

The application of equation (4.2) to the mechanism shown in Fig. 4.13 is very simple. An additional feature has been added to the velocity diagram. The head of each vector has been labelled with a lower-case letter, for example the head of $\dot{Q}$ is labelled q. Points A and B on the linkage have zero velocity, so the labels a and b are placed at the origin of the velocity diagram. The direction-line on the velocity diagram has also been added, with shading to identify it, although this is not usually shown, it being assumed that velocities are measured in the same direction on the velocity diagram as in the linkage. The heads of the vectors in the velocity diagram are called *images* of the corresponding points in the linkage – more precisely, velocity images.

There are mechanisms which cannot be analysed by the methods described here (an example will be given in a later chapter). In addition all graphical methods suffer from the disadvantage of analysing the mechanism in one position only. For an investigation of velocities throughout the cycle it is necessary to draw a series of diagrams; this is tedious, and becomes more so if accelerations must also be investigated. Analytical methods which can deal with any mechanism, and which can be computerised to give rapid analysis of the whole cycle, will be described later in this book.

4.5   Contour cams
These are mechanisms in which motion is transmitted by polode rolling, either directly as in Fig. 4.14, or by using the polodes as pitch surfaces on which gear-teeth are cut, for example elliptic and other non-circular gears.

Fig. 4.14. Design of contour cams.

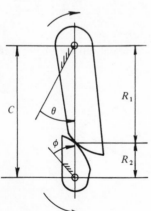

The profiles of contour cams can be constructed graphically as smooth curves drawn through poles. But these mechanisms are mostly used when very high accuracy is required, say in a computing mechanism, so it is advisable to calculate the profiles. This is done as follows (Fig. 4.14). The profile of each cam is specified by polar co-ordinates (R, θ) measured from an initial line in the cam, shown with shading. The centre-distance C and a relationship $\phi = f(\theta)$ between input and output angles are specified. Then, because the point of contact lies on the line of centres (the Aronhold–Kennedy theorem),

$$R_1 + R_2 = C \tag{4.3}$$

and because there must be no slip at the point of contact

$$R_1 \delta\theta = R_2 \delta\phi$$

and hence, in the limit,

$$R_1 - \frac{d\phi}{d\theta} R_2 = 0 \tag{4.4}$$

Solving equations (4.3) and (4.4) simultaneously we obtain the design formulae

$$R_1 = \frac{C\phi'}{1 + \phi'}$$

$$\tag{4.5}$$

$$R_2 = \frac{C}{1 + \phi'}$$

where $\phi' = d\phi/d\theta$. Since ϕ' is a function of θ, the first equation gives R_1 as a function of θ, and specifies the profile of the upper cam in the figure. By writing ϕ' as a function of θ we make the second equation give R_2 as a function of ϕ, and thus specify the profile of the other cam. Probably the easiest way to understand the process is to verify the example shown in Fig. 4.14, which generates $\phi = 0.1\theta^2$ for $10° \leqslant \theta \leqslant 30°$.

4.6   Epicyclic gearing

The instantaneous-centre method of velocity analysis is particularly suited to epicyclic gearing because all the centres have obvious locations. These are shown in Fig. 4.15 for a common type of train. Only one planet is shown since the others behave similarly and their removal would not affect the overall kinematics of the train.

Suppose now that we fix the annulus (link 1) and drive the sun gear (link 2). Consider the motion of the planet, indicated in Fig. 4.16. Since we are driving the sun at a known speed the velocity of the point of contact P_{24} between sun and planet can be calculated and represented by a line drawn to a convenient scale. Since the annulus is fixed P_{14} is at rest and therefore, by the construction in the figure, we can find the velocity of the planet centre P_{34}. But this is also a point on the arm, link 3, and thus we can find the angular velocity of the arm.

4.7 Finding the pole by complex numbers

The account of the instantaneous centre given in this chapter relies on physical and geometrical intuition, and its main purpose is to provide a picture of what is really going on in a mechanism at any instant. It is interesting to look at this most important topic from another viewpoint.

In Fig. 4.17 we show a link in general plane motion, its position at any instant being specified by the position of a marked point Z on the link and the inclination θ of a line fixed in the link. A point P has a position vector $\mathbf{P}$ measured in the

Fig. 4.15. Instantaneous centres of an epicyclic gear-train.

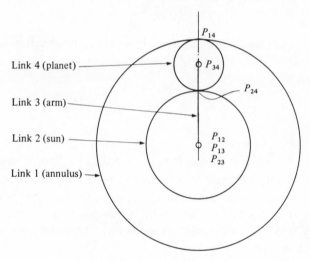

Fig. 4.16. Finding angular velocities in an epicyclic train.

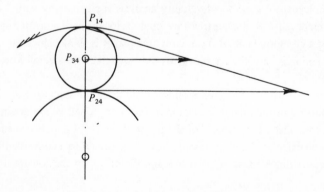

Fig. 4.17. Use of complex numbers to locate the pole.

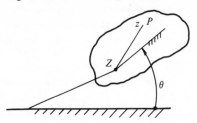

fixed link and position vector z measured in the moving link. **P**, **Z** and z will be treated as complex numbers; **P** and **Z** are variables, while z is a constant. The velocity $\dot{\mathbf{Z}}$ of the point Z, and the angular velocity $\dot{\theta}$, are supposed given. We will calculate the velocity $\dot{\mathbf{P}}$, and find the location of P so that $\dot{\mathbf{P}} = 0$, i.e. we will prove the existence of a pole and locate it in both links. We have

$$\mathbf{P} = \mathbf{Z} + z e^{i\theta} \tag{4.6}$$

and therefore

$$\dot{\mathbf{P}} = \dot{\mathbf{Z}} + i\dot{\theta} z e^{i\theta}$$

Putting $\dot{\mathbf{P}} = 0$ and solving the last equation for z we have

$$z = - i(\dot{\mathbf{Z}}/\dot{\theta}) e^{-i\theta}$$
$$= - i(d\mathbf{Z}/d\theta) e^{-i\theta}$$

which locates the pole in the moving link. Substituting into equation (4.6) we also have

$$\mathbf{P} = \mathbf{Z} - i(d\mathbf{Z}/d\theta)$$

which locates the pole in the fixed link. Incidentally this last result also shows that the pole lies on a line perpendicular to $\dot{\mathbf{Z}}$, i.e. on a path-normal.

Exercises 4

4.1. The linkage shown in Fig. 4.18 is driven by rotating the left-hand crank counter-clockwise at 100 revs/min.

Fig. 4.18.

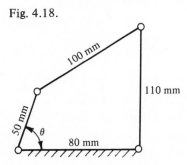

EXERCISES 4(*contd.*)

(*a*) When $\theta = 60°$ find the angular velocity of the coupler by the instan-taneous-centre method and also by means of a velocity diagram. Com-pare the results of the two methods. What sort of percentage accuracy would you hope to achieve in the careful graphical analysis of velocities, and to how many figures should results usually be quoted?

(*b*) Find the angular velocity of the coupler and the right-hand crank, and the velocity of the mid-point of the coupler, when $\theta = 90°$. If the left-hand coupler pivot consists of a 12 mm diameter plain bearing, calculate the rubbing speed in m/s.

4.2. In the mechanism shown in Fig. 4.19 the small gear-wheel is free to rotate about the fixed pivot and the large gear-sector is rigidly attached to the coupler. Locate all the instantaneous centres. Hence or otherwise find the angular velocity of the small gear-wheel when $\theta = 60°$ and the right-hand crank is driven at an angular velocity of 100 revs/min clock-wise.

Fig. 4.19.

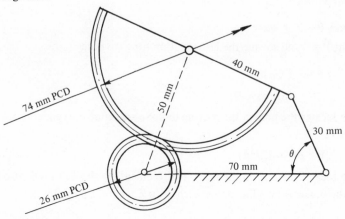

4.3. Fig. 4.20 shows two inversions of the slider–crank mechanism. Find, for each of them, the sliding velocity of link 4 relative to link 1 when link 2 is rotated at 100 revs/min counter-clockwise and $\theta = 45°$.

Fig. 4.20.

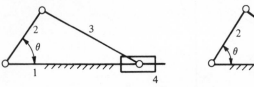

Link 2: 50 mm
Link 3: 75 mm

4.4. For the mechanism shown in Fig. 4.21 find the rubbing speed of the slider on the coupler, in m/s, when the left-hand crank is rotated at 50 revs/min clockwise and $\theta = 80°$. Would the result be the same if the rotation of the left-hand crank were counter-clockwise? Give reasons for your answer.

Fig. 4.21.

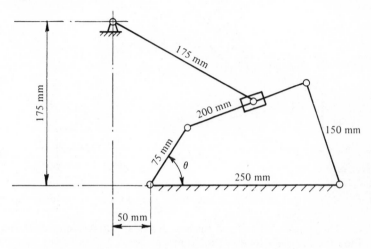

4.5. For the 6-bar linkage shown in Fig. 4.22 find the angular velocity of link 4 when link 3 has a clockwise angular velocity of 10 revs/min and $\theta = 90°$.

Fig. 4.22.

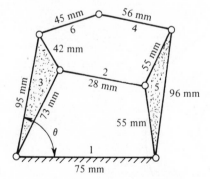

4.6. In the mechanism shown in Fig. 4.23 link 2 rotates at 100 revs/min counter-clockwise. Find the velocity of the slider, link 6, in m/s when $\theta = 65°$

(*a*) by a velocity diagram, and (*b*) by instantaneous centres.

Fig. 4.23.

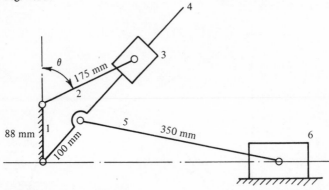

4.7. Fig. 4.24 shows an epicyclic gear-train, the tooth-numbers being shown next to the gears.

(*a*) The annulus is fixed and the arm driven at 100 revs/min clockwise. Find the angular velocity of the sun.

(*b*) The sun is fixed and the annulus driven at 10 revs/min clockwise. Find the angular velocity of the arm.

(*c*) The sun is driven at 100 revs/min counter-clockwise. At what speed must the annulus be driven, and in what direction, in order that the arm shall be stationary; and what is the angular velocity of the planet?

Fig. 4.24.

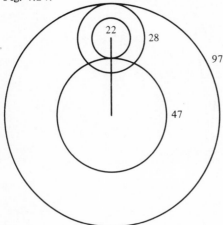

4.8. In the epicyclic train shown in Fig. 4.25 shaft A is driven at 100 revs/
min counter-clockwise viewed in the direction of the arrow. Find the
angular velocities of the arm and shaft B.

Fig. 4.25.

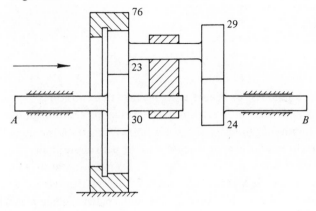

5   AN INTRODUCTION TO CURVATURE THEORY

In the synthesis techniques considered so far we take three positions of a point
on a moving link and locate the centre of the circle that passes through them.
An alternative approach is to take a single position of the moving point and find
the circle that is the best local approximation to the path of the point. This circle
is called the circle of curvature. Since the link rotates instantaneously about the
pole it might at first be thought that the pole is the centre of curvature. But this
is not so, because the pole itself is moving. The actual paths of some points on
the coupler of a 4-bar linkage are shown in Fig. 5.1, where the coupler is extended
into a large rectangular link that covers the whole mechanism. The point P' which
coincides with the pole moves locally on a cycloidal path. Any other point has a
path whose tangent is perpendicular to the line joining that point to the pole. The
coupler pivots, of course, move on circular arcs and have the fixed pivots as
permanent centres of curvature. There is a circle passing through the pole, called

Fig. 5.1. Paths of points on the coupler of a 4-bar linkage.

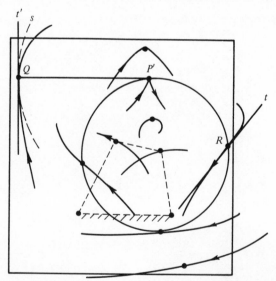

the *inflexion circle*, and every point on this circle (except the pole itself) is at a point of inflexion on its path. Suppose R is such a point and t the tangent to the path of R at the inflexion point. Then, for some distance on either side of the inflexion point, the path of R will lie quite close to t. Such points provide good approximations to straight-line motion. In all cases of plane motion, as well as in this example, there is an inflexion circle with this property. The path of a point Q, with its tangent t' and circle of curvature s, are shown in the figure. It can be seen that s is locally much closer to the point-path than t', and also that the point-path crosses the circle of curvature at the point of contact. At a point of inflexion, the radius of curvature is infinite; the circle of curvature becomes a tangent.

5.1   Hartmann's construction

The point-path and the circle of curvature have a common tangent at the point of contact. Hence the centre of curvature A_0 will lie on the line joining the moving point A to the pole P. This, like any line through the pole, is called a *ray*. Since the pole must always lie on the ray, a point on the ray coincident with the pole must be moving with a velocity equal to the component of pole velocity at right angles to the ray. Hence, if we know the pole velocity, we have the simple construction for finding the centre of curvature shown in Fig. 5.2. Draw a ray through the moving point A and the pole P. Draw the velocity vector AR of A.

Fig. 5.2. Hartmann's construction for finding the centre of path curvature A_0 for a moving point A.

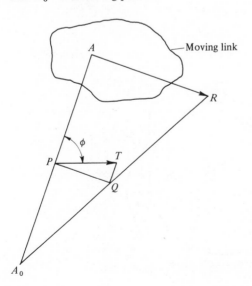

Draw a line through P parallel to AR, and drop a perpendicular onto this line from the head of the pole velocity PT to cut this line at Q. Now draw a line through R and Q. This will cut the ray at the required centre of curvature A_0.

Hartmann's construction depends on knowing the pole velocity, and a method of finding this will be given below. It will appear, however, that in certain cases (such as the 4-bar linkage and slider–crank), the pole velocity is not required.

5.2   The inflexion circle

Suppose the ray PA makes an angle ϕ with the pole velocity (Fig. 5.2). Then by similar triangles we have:

$$\frac{A_0 A}{AR} = \frac{AP}{PQ}$$

so

$$\frac{AA}{AP} = \frac{AR}{PQ}$$

but $A_0 A = A_0 P + PA$, $PQ = PT \sin \phi$, and $AR = \omega PA$, where ω is the instantaneous angular velocity of the link about the pole P. Hence the second equation can be written as.

$$\left(\frac{1}{A_0 P} + \frac{1}{PA}\right) \sin \phi = \frac{\omega}{PT} \tag{5.1}$$

where the right-hand side is a constant for any particular position of the link, i.e. its value does not depend on the choice of the moving point A. Now if J is a point of inflexion, $J'P = \infty$, so equation (5.1) reduces to

$$PJ = \frac{PT}{\omega} \sin \phi \tag{5.2}$$

Fig. 5.3. A construction relating a moving point A, its centre of path curvature A_0 and a point of inflexion J.

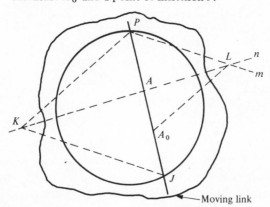

and this is the equation in polar co-ordinates (PJ, ϕ) of a circle passing through P, at which point it is tangent to the pole velocity, and of diameter PT/ω. Thus all points of inflexion lie on a circle in any type of plane motion.

Suppose a point of inflexion J and any other point A both lie on the same ray (Fig. 5.3). Let A_0 be the centre of curvature of A. Multiplying equations (5.1) and (5.2) together gives $PJ[(1/A_0P) + (1/PA)] = 1$. But $PJ = PA + AJ = PA - JA$ (since direction must be taken into account), and similarly $A_0P = A_0A - PA$. Thus

$$A_0A = (PA)^2/JA \qquad\qquad (5.3)$$

In this equation the moving point, its centre of curvature, the pole, and a point of inflexion are all related without any mention of velocities. We could solve the equation with a slide-rule, remembering that A_0A must be measured in the same direction as JA, but the graphical solution shown by broken lines in the figure is simpler and more reliable, since directions come right automatically. There are two cases to consider:

Construction A. Given a point A, the pole P and a point of inflexion J, to find the centre of curvature A_0.

Draw any line m through P, and any line n through A. Let L be the point of intersection of m and n. Draw a parallel to m through J, to cut n at K. Now draw a parallel to PK through L. This cuts the ray in the required point A_0.

Construction B. Given the pole P, a moving point A and its centre of curvature A_0, to find the point of inflexion J on the ray through A.

We draw the same diagram as in Construction A, but in a different sequence. Draw the lines m and n, and mark their intersection L as before. Join L to A_0. Draw a line through P parallel to LA_0, to cut n at K. Now draw a parallel to m through K. This cuts the ray at the required point of inflexion J.

These constructions depend only on drawing parallel lines, which can easily be done accurately; and a check can be made by repeating the construction with a new pair of lines m and n. A proof that the construction solves equation (5.3) can easily be obtained by considering the similar triangles in the figure.

5.3 The slider–crank and the 4-bar linkage

The constructions described above can be used to obtain the inflexion circle, and the centre of curvature of any coupler-point, on either the slider–crank or the 4-bar linkage. The slider–crank (Fig. 5.4) is the simpler case. We find the pole P by the method described in Chapter 4. The cross-head B moves in a straight line, and so it is always at a point of inflexion on its path. The pivot A has a centre of curvature at A_0, the fixed pivot. We therefore use Construction B to locate the point J. Now the inflexion circle must pass through P, B and J, and since three points determine a circle it can be drawn. The curvature of any other coupler-point can then be found by Construction A. In this case the line n has been drawn as a continuation of the link AB, which is permissible since n can be *any* line through A.

The 4-bar linkage (Fig. 5.5) is solved using Construction B twice. A single line n, drawn through A and B, and a single line m through the pole P serve for both applications of the construction. As in the case of the slider–crank we use the fact that points A and B have known centres of curvature A_0 and B_0 respectively. The procedure is as follows. Find the pole P as in Chapter 4. Draw a line n through AB, and any line m through P to cut n at L. Draw a line parallel to $A_0 L$ through P to cut n at KA. Draw a line parallel to m through KA to cut $A_0 A$ in J_A. Now draw a line parallel to $B_0 L$ through P to cut n in K_B. Draw a line parallel to m through K_B to cut $B_0 B$ in J_B. The inflexion circle can now be drawn through the three points, P, J_A and J_B. The curvature of any coupler-point can now be found by Construction A.

Fig. 5.4. A construction for the inflexion circle of a slider–crank mechanism.

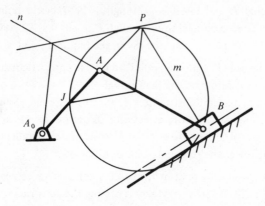

Fig. 5.5. A construction for the inflexion circle of a 4-bar linkage.

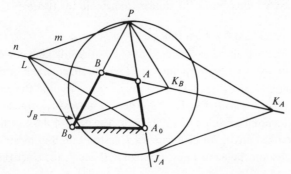

5.4   Applications

Most of the applications of curvature theory depend on some further constructions and will be considered in Chapter 6. There is, however, a simple and very useful application that can be explained here, namely finding a point on the coupler of a 4-bar linkage that moves in a good approximation to a straight line and in some required direction. A specific example, the design of a level-luffing crane, is shown in Fig. 5.6.

The action of a crane in moving its load in a horizontal direction is called luffing. When this occurs it is important that there should be as little vertical motion of the load as possible. Any vertical motion will require unnecessary work to be done on the load, wasting power and increasing the size and cost of the motor, brake and gearing used for luffing. Ideally the jib pulley should move in a straight horizontal line, with a slight rise at each end of its travel to assist braking. A good approximation can be obtained as follows:

The main part of the crane is a 4-bar linkage, designed to give adequate height and to rock through the required horizontal travel. We then seek a point on the coupler which moves approximately in a horizontal straight line. Find the pole P and draw the inflexion circle s (only part of the circle is shown in the figure because in this example its diameter is quite large). Now draw a vertical line through P to cut the inflexion circle in some point Q. Then Q will, at that instant, move perpendicularly to PQ, i.e. in a horizontal direction, and will have a point of inflexion. The actual path of Q is shown in the figure. The construction should

Fig. 5.6. Use of the inflexion circle in the design of a level-luffing crane. This construction can be used in many problems where it is necessary to obtain approximate straight-line motion in a specified direction from a linkage.

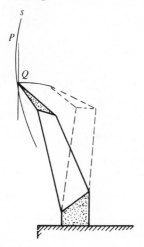

be carried out with the crane drawn at some intermediate position in its luffing motion, and it will usually be necessary to carry out a few trials in order to get a satisfactory design.

5.5   Finding the pole velocity

The constructions based on Fig. 5.3 can only be used if we can locate two moving points in the mechanism whose centres of curvature are known (points of inflexion, such as the cross-head in a slider–crank, being included). When this information is not available we can fall back on Hartmann's construction, for which we require a knowledge of the pole velocity. Now the pole is always located at the intersection of two lines and we are therefore able to find its velocity. The method is general, but is illustrated here by a 4-bar linkage (Fig. 5.7). Suppose the two lines that locate the pole are materialised as slotted links trapping a small cylinder, as in the upper diagram. Then the pole will be at the centre

Fig. 5.7. Finding the pole velocity, pole tangent and pole normal.

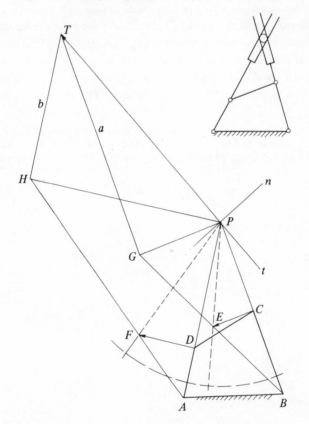

of this cylinder, the velocity of which will be the pole velocity. We find it by the construction shown in the lower diagram.

First locate the pole P. Given the velocity of either of the coupler pivots C or D, find the velocity of the other one by the tetta-angle construction shown dashed. Draw a line through B and E. Drop a perpendicular to BP from P to cut this line in the point G. Then PG is the velocity of a point on the link BC at the pole, and hence it is the component of pole velocity normal to the link BC. But there must also be a component along the link, so we draw a line a perpendicular to PG through G. The head of the pole velocity vector must lie somewhere on this line.

We then draw a line through A and F, and drop a perpendicular to AP from P to cut this line at H. Draw a line b perpendicular to PH through H. Then, by the same argument as applies to the line a, the head of the pole velocity vector must also lie on b. Hence it lies at point T, the intersection of a and b. Thus the pole velocity is PT. The line t along the pole velocity is the common tangent to the polodes at the point of contact and is called the *pole tangent*. The line n perpendicular to t and passing through P is called the *pole normal*, or *principal ray*.

It may happen that we do not know the velocity of any point in the mechanism. Since curvature is a purely geometrical property, and has nothing to do with the actual speed at which the mechanism runs, we simply draw any convenient line CE perpendicular to BC at C, and carry out the construction as above. Then, *using the same tetta-angle*, we draw the inflexion circle by the construction shown in Fig. 5.8. We have already found the pole P, the pole tangent PT and the pole normal n. Draw a line through P at the angle tetta to n, making sure

Fig. 5.8. Finding the inflexion pole and the inflexion circle when the pole velocity is known.

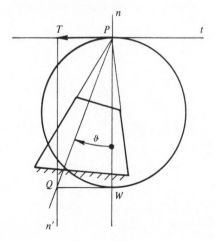

that tetta is taken in the same sense as used for the pole velocity construction (in this example tetta is clockwise). Now draw a line n' parallel to n through T, to cut the tetta-line at Q. Finally draw a line parallel to PT through Q to cut n at W.

It can be seen that, according to Hartmann's construction, W will be a point of inflexion, for WQ is the velocity of W and TQ is parallel to PW. Hence W (which is called the *inflexion pole*) lies on the inflexion circle. Now the pole tangent is a tangent to the inflexion circle at P, so PW is the diameter of the circle. Hence the inflexion circle can be drawn, and we can now use the construction shown in Fig. 5.3 to find the curvature of any point-path. In practice, of course, the construction for W is performed on the same drawing as the construction by which we find the pole velocity; they are drawn separately here to make the diagrams clearer.

This method of finding the inflexion circle requires us to draw equal angles (the tetta-angles) and lines perpendicular to each other, and is therefore less accurate (as well as lengthier) than the construction of Fig. 5.3, which should always be preferred when it is possible to use it.

5.6   The Euler–Savary equation and Cardan motion

The constant ω/PT, where ω is the angular velocity of the link about the pole and PT is the pole velocity, appears in equation (5.1). In further working we eliminated this constant and thus avoided having to find its value. It can, however, easily be found. Consider Fig. 5.9, where the polodes have been replaced by their circles of curvature at the point of contact. We adopt a labelling convention that is important in more advanced work. The fixed and moving polodes have radii r_1 and r_2 respectively; both these values are taken as positive if (as in

Fig. 5.9. Relation between pole velocity and angular velocity of the moving polode.

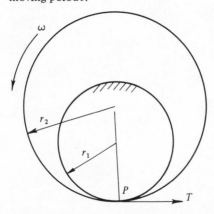

the figure) the centres are *above* the pole tangent; if below they are taken as negative. The angular velocity ω is taken as positive for the counter-clockwise sense, and the pole velocity PT is taken as positive if it is directed to the right. Imagine the centre of the moving polode joined to that of the fixed polode by a link, as in epicyclic gearing. Then, since this link must pass through the pole at all times, it is an easy exercise (left for the reader) to show that

$$PT = \frac{\omega r_1 r_2}{r_2 - r_1}$$

and hence

$$\frac{\omega}{PT} = \frac{1}{r_1} - \frac{1}{r_2} \tag{5.4}$$

Using our labelling convention this equation is true for all values of r_1, r_2 and ω. We now substitute this result into equation (5.1), obtaining the *Euler–Savary equation*

$$\left(\frac{1}{A_0 P} + \frac{1}{PA}\right) \sin \phi = \frac{1}{r_1} - \frac{1}{r_2} \tag{5.5}$$

In this equation the moving point and its centre of curvature are related solely by the polode curvatures, and in most books on kinematics it is the practice to prove this equation first by purely geometrical arguments, and then deduce the rest of the curvature theory from it. A different order of exposition has been followed here in the hope of making the theory easier.

An interesting thing happens if we make $r_1 = 2r_2$, giving the case shown in Fig. 5.10, where the moving polode rolls internally on a fixed polode of twice

Fig. 5.10. Cardan motion. If the moving polode rolls internally on a fixed polode of twice its radius, the moving polode is also the inflexion circle.

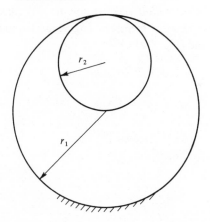

Fig. 5.11. Use of Cardan motion in an engine mechanism.

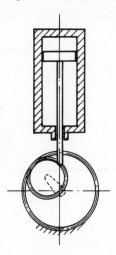

the radius. From equation (5.4) we now have $\omega/PT = -1/2r_2$. Recalling equation (5.2) we find that the diameter of the inflexion circle must be $-2r_2$. Now it can be shown (but the argument is beyond the scope of this book) that a negative value for the diameter of the inflexion circle must, with our labelling conventions, mean that the circle is above the pole tangent; while a positive value means that it is below. The point does not arise in the graphical constructions given in the rest of this chapter, but it does if we use analytical methods. In this case it means that the moving polode is itself the inflexion circle. When this occurs the moving link is said to be in a *Cardan position*.

If both the polodes are actually circles, say the pitch circles of mating gear-wheels, it follows that every point on the moving polode will *always* be at a point of inflexion on its path. Now this is only possible if the entire path is a straight line. This phenomenon is called *Cardan motion* and the polodes are called Cardan's circles. Cardan motion has been used in a steam-engine mechanism for straight-line guidance of the piston-rod (Fig. 5.11).

Exercises 5

5.1. For the 4-bar linkage shown in Fig. 5.12 find the centre of curvature
for the path of point *C* when θ = 43°. Draw the circle of curvature and
plot the complete path of *C*.

Fig. 5.12.

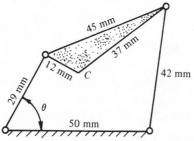

5.2. For the 4-bar linkage shown in Fig. 5.13 draw the inflexion circle and
plot the complete path of the coupler-point which, in the position shown,
is coincident with the instantaneous centre.

Fig. 5.13.

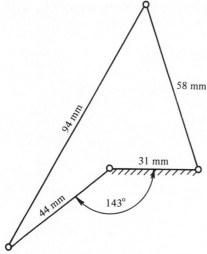

5.3. By considering the trammel method of drawing an ellipse show how the
curvature can be found for any given point on the ellipse. Comment on
the usual circular-arc construction for the approximate drawing of ellip-
ses.

5.4. Show, from the theory of Cardan motion, that straight-line motion can
be generated using *external* gears only. Hence devise a lathe attachment
for generating a flat on a round bar.

6 FURTHER NOTES ON CURVATURE THEORY

Basic curvature theory, with proofs of the various equations and constructions, was described in the previous chapter. Some further constructions are needed for convenience in practical applications, and will be given here without proofs in order to keep this chapter reasonably brief.

6.1 Bobillier's theorem

Suppose that on a moving link (Fig. 6.1(a)) two points A and B have known centres of curvature A_0 and B_0 respectively. Draw the lines A_0A and B_0B to intersect at the pole P. Draw the lines AB and A_0B_0 to intersect at a point Q. The line PQ is called the *collineation axis*, or centre-join, for the rays A_0A and

Fig. 6.1. Bobillier's theorem.

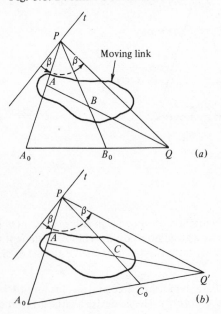

B_0B. We will call the ray furthest from the collineation axis the first ray, and the ray next to the collineation axis the second ray. Bobillier's theorem states that

> *The angle from the pole tangent to the first ray equals the angle from the second ray to the collineation axis.*

We can therefore draw the pole tangent t, as in Fig. 6.1(a), where the two equal angles are each labelled β. Now suppose we have a third point C on the moving link, as in Fig. 6.1(b), and we wish to find its centre of path curvature. We draw the ray PC, and lay off a new collineation axis at the same angle β. We now work from the original first ray. Draw a line through AC to cut the new collineation axis at Q'. Draw a line through A_0Q' to cut the new ray at C_0. Then C_0 is the required centre of curvature for C.

In these constructions it must always be remembered that the angle β must be drawn in a definite direction, i.e. *from* the second ray *to* the collineation axis.

In a practical case the two constructions shown in Fig. 6.1 would be drawn on the same diagram, but they are separated here for the sake of clarity. This method of finding the centre of curvature can be simplified still further if we take account of the fact that we do not need to draw the pole tangent. An application to the 4-bar linkage is shown in Fig. 6.2. We wish to find the centre of curvature for the coupler-point C. Construct P, Q and the collineation axis for the two cranks as shown in the figure, and mark the angle β. The rest of the construction is shown in broken lines. Draw the ray PC and lay off the new collineation axis at the angle β. Join AC to give a line cutting the new collineation axis at Q'. Now join A_0Q' by a line which will cut the ray PC at C_0, the required centre.

It sometimes happens that a particular location in the machine we are designing will serve conveniently as a fixed centre, and we wish to find the appropriate moving point. In that case the above construction can easily be reversed. We proceed as before until we have drawn the new collineation axis. Then draw

Fig. 6.2. Use of Bobillier's theorem to find the centre of path curvature C_0 for the point C.

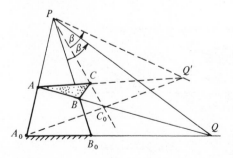

a line through A_0 and the given centre C_0, to cut the new axis in Q'. Finally, join Q' to A by a line which will cut the new ray at the required point C.

An interesting application of Bobillier's theorem is given by K. H. Hunt, *Mechanisms and Motion*, London: English Universities Press, 1959. The mechanism shown in thick lines in Fig. 6.3 is proportioned so that $AB = BD = B_0B$. It is called an *isosceles slider-crank*, and has a number of useful properties, amongst them the fact that the point D traces an exact straight line (this is easily proved by elementary geometry). This mechanism can be used in an engine indicator, where the piston works against a stiff spring and has the very short stroke shown in the figure. When so used the point D covers the distance between the short horizontal lines along B_0A_0 and thus magnifies the piston stroke. A pencil mounted at D marks a trace on a revolving drum. Much of the working load on the mechanism is provided by the inertia of the link B_0B, which unfortunately is rather large and is driven at a poor transmission angle. Also the pivot B_0 is inconveniently located, making it difficult to mount the drum.

Here we make a typical use of curvature theory, namely the modification for practical purposes of an unsatisfactory basic design. We start by choosing a new, and more convenient, location for the fixed pivot at the point C_0. We must now find the corresponding moving point C on the link AB, by applying Bobillier's theorem. The centre of curvature for A lies at infinity on a line perpendicular to the cylinder axis, and this line AA_0 will be the first ray. It is intersected by the second ray B_0B at the pole P. The line B_0A_0 is parallel to AA_0 (since A_0 is at infinity) and is intersected by AB at the point D. The line PD is the collineation axis, and we now have the angle β.

The rest of the construction is shown in broken lines. We draw the ray PC_0 and lay off the new collineation axis at the angle β. Now draw a line through C_0

Fig. 6.3. Use of Bobillier's theorem in the design of an engine indicator.

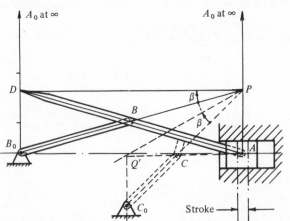

parallel to AA_0. This is the line C_0A_0 and it intersects the new collineation axis at Q'. Finally join AQ' by a line which intersects the ray PC_0 in the required point C. The link B_0B is, of course, removed, and we use instead the link C_0C shown by broken lines. This is a shorter link, with less inertia, working at a better transmission angle, and with a fixed pivot located more conveniently. The re-designed mechanism traces a path which is a very good approximation to a straight line. The new path is also slightly longer at the lower end, thus increasing the indicator's magnification.

Bobillier constructions are very simple and quick to use, and introduce only a few extra lines into the drawing. Hence they are useful in preliminary design, particularly if several trials need to be made. Success depends on the accuracy with which the angle β is transferred from one part of the drawing to another, and perhaps it is unnecessary to point out that this should be done by stepping off equal arcs with dividers on a single large circle rather than by using a protractor or adjustable square. When high accuracy is needed, say in the final checking of a design, it is better to use the constructions given in Chapter 5.

6.2 Stationary curvature

All the constructions in curvature theory apply only to a particular position of the mechanism. As the moving point continues on its path, the curvature of that path will change. In doing so it will pass through maximum and minimum values, at which its rate of change will be zero. These values are therefore called stationary points. Near a stationary point the rate of change of curvature will be small and thus the path will be an exceptionally good approximation to a circle. Points of stationary curvature are therefore of value in mechanism design, and a construction for finding these points in the case of a 4-bar linkage will now be described.

The construction, illustrated in Fig. 6.4, is shown in two stages which in practice would both be carried out on the same drawing. Referring first to Fig. 6.4(a), draw the pole, pole tangent t, and pole normal n by any of the constructions already given. Now draw a line through A in the direction of the velocity of A, i.e. perpendicular to A_0A, to cut t at A_t and n at A_n. Complete the rectangle A_tPA_n to obtain the point A_g. Working now with point B, perform a similar construction to get point B_g. A line g is drawn through Ag and B_g.

The rest of the construction is shown in Fig. 6.4(b). Take any point R_g on g, and from it drop perpendiculars onto t to give R_t and n to give R_n. Join R_t to R_n, and drop a perpendicular onto this line from P to give point R. Then R is a point of stationary curvature.

By choosing a succession of points on g and repeating this last part of the construction we obtain as many points as required and draw through them the locus, which is known as the cubic of stationary curvature, since it has a cubic algebraic equation. The curve passes through the coupler pivots (which are of course points of stationary curvature because they lie on circular arcs) and the pole, and is tangent at the pole to both the pole normal and the pole tangent.

The construction applies not only to the 4-bar, but to any case in which we know that two points on a moving link have stationary curvature. The extension

Fig. 6.4. Construction for the cubic of stationary curvature.

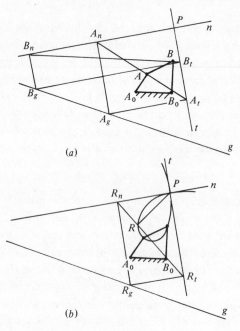

(a)

(b)

Fig. 6.5. Ball's point R is at the intersection of the inflexion circle with the cubic of stationary curvature.

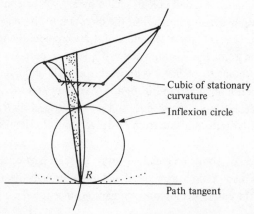

Cubic of stationary curvature

Inflexion circle

R

Path tangent

to a slider–crank mechanism is obvious, since the cross-head pivot has stationary curvature and a known direction of motion along the guide axis.

A point of particular importance is that at which the cubic intersects the inflexion circle. This is known as *Ball's point*, and is a point of *undulation*, at which the path is an even closer approximation to a straight line than at an ordinary point of inflexion. Moreover, while at an inflexion point the path crosses its tangent, at an undulation point it lies entirely on one side of it. The construction for Ball's point R is shown in Fig. 6.5, and the path of this point is shown dotted. It can be seen that the path practically coincides with the tangent for a distance (in this example) equal to about twice the length of the smaller crank. In this way rotary motion can be converted into approximate straight-line motion in circumstances where a slider–crank mechanism would have impossibly bad transmission angles.

It sometimes happens that this construction cannot be used because a point of intersection, or the line g, lies off the paper. In such a case the curve can be plotted from its equation in polar co-ordinates, which is

$$\frac{1}{r} = \frac{1}{M \sin \phi} + \frac{1}{N \cos \phi} \tag{6.1}$$

where r and ϕ are polar co-ordinates with the pole as origin and the pole tangent as initial line. We can find the values of the constants M and N by measuring the values of r and ϕ at two known points on the curve, for example the coupler pivots in the case of a 4-bar linkage.

6.3 Equilibrium of rolling bodies

Points on the pole normal inside the inflexion circle have paths which are convex to the pole. Hence a rolling body will be in stable equilibrium on a flat surface if the mass centre lies inside the inflexion circle. This principle has been used in the design of toys that rock but do not fall over, and also in an 'unspillable' salt-cellar, as shown in Fig. 6.6. The mass centre M lies on the curved path shown by

Fig. 6.6. Equilibrium of rolling bodies. The body will be in stable equilibrium if its mass centre lies inside the inflexion circle.

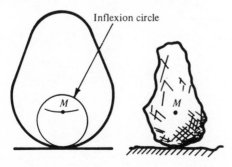

a thin line, and if the salt-cellar is rocked M will rise, thus increasing the potential energy. Hence the salt-cellar will be in a stable equilibrium when it is vertical, this being the position with the least potential energy. Nearer the pole the path is curved more sharply, meaning that as the body rocks the restoring force will increase more rapidly. Hence the lower the mass centre the more rapidly will the body vibrate when it is rocked. The applications of this theory are to toys and novelties, and also to the 'rocking stones' sometimes found in mountainous country (Fig. 6.6). Such a stone, despite its weight, can be rocked but will return to the vertical – if the displacement is not too large.

Exercises 6

6.1. For the 4-bar linkage shown in Fig. 6.7 locate Ball's point and plot its complete path. For what fraction of the mechanism's cycle does this path coincide (within graphical accuracy) with a straight line?

Fig. 6.7.

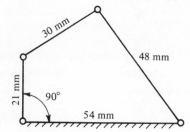

6.2. Revise your solution of question 1.5 by considering the cubic of stationary curvature.

7 COMPLEX LINKAGES

7.1 The need for complex linkages

Most sorts of output motion required in practice can be obtained from gears, cams, or the simple linkages described in earlier chapters. Complex linkages are, however, sometimes necessary in order to obtain an elaborate output motion (for example in textile machinery) under conditions of speed and load which would soon wear out a cam; 10-bar and even 12-bar linkages have proved commercially practicable in service. It also sometimes happens that although a 4-bar or 6-bar would give the required motion, extra links must be inserted to transmit that motion over a distance. In the machine shown in cross-section in Fig. 7.1,

Fig. 7.1. The need to transmit motion over a distance can increase the number of links in a mechanism.

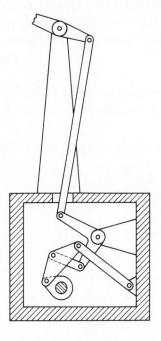

for example, the 6-bar linkage must be kept small enough to fit in the crank-case and motion must be transferred via a connecting-rod to an extra rocker at the work station. In this way a 6-bar grows into an 8-bar.

When it appears inevitable that more than four bars are going to be used further design alternatives present themselves because of the large variety of ways in which the bars can be connected together. The number of choices open to the designer becomes, as will be shown in this chapter, much larger than is generally realised. Let us take a definite example (Fig. 7.2). We wish to mount an implement on the back of a tractor, using a rotary actuator to raise and lower the implement while keeping it approximately parallel to the ground. It is an obvious first step to mount the implement on the coupler of a 4-bar linkage, of which the tractor itself is the fixed link. Now the rotary actuator could be used to rotate one of the cranks of this linkage, but the small rotations required to raise and lower the implement would use only part of the actuator's available stroke, and would require fairly sensitive control gear as well as a high actuator torque. We therefore mount the actuator separately on the frame, and introduce two small additional links. Thus we arrive at a 6-bar linkage.

Any competent designer, reasoning in much the same way, would probably arrive at this result. It might then occur to him to stop and consider if there are any alternative ways of doing the job with the same number of links. After trying this problem on a number of designers it appears that about three or four alternatives can be produced by intuitive inventing, usually after a good deal of effort. In fact, however, there are no less than 24 alternative ways of doing the job using only pivoted links. We will show how these alternatives can be discovered.

7.2 Systematics of linkages

We now use some of the ideas and definitions given in Chapter 1. For a pivoted linkage the Chebyshev–Grübler formula takes the simple form

$$M = 3(n - 1) - 2j \tag{7.1}$$

where M is the mobility, n the total number of links, and j the total number of

Fig. 7.2. Use of a linkage to mount an implement on the back of a tractor.

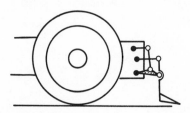

pivots. We limit ourselves here to cases where $M = 1$, so equation (7.1) can be re-written as

$$3n - 2j = 4 \tag{7.2}$$

Now there will be k elements on each link of order k, and provided we use only simple hinges, two elements will be needed for each joint. Thus

$$2j = 2n_2 + 3n_3 + 4n_4 + \ldots \tag{7.3}$$

where n_k is the number of links of order k. Also the total number of links is ob-viously the sum of the numbers of links of each order, so

$$n_2 + n_3 + n_4 + \ldots = n \tag{7.4}$$

We now substitute equation (7.2) into equation (7.3) to obtain

$$2n_2 + 3n_3 + 4n_4 + \ldots = 3n - 4 \tag{7.5}$$

and by solving equations (7.4) and (7.5) we can find what kinds of links and how many of them make up an n-bar linkage. This is easily shown by an example.

Suppose, as in the example considered here, $n = 6$. Then the equations are

$$2n_2 + 3n_3 = 14$$

$$n_2 + n_3 = 6$$

Fig. 7.3. A 'kit' (*a*) of four binary and two ternary links can be as-sembled to give either of two kinematic chains (*b*) or (*c*). From (*b*), by introducing double hinges, we obtain two further chains (*d*) and (*e*). The same result would be obtained by introducing double hinges into chain (*c*).

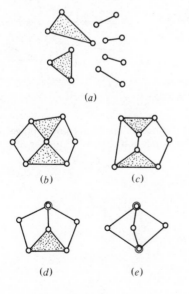

(*a*)

(*b*)   (*c*)

(*d*)   (*e*)

of which the solutions are $n_3 = 2$, $n_2 = 4$. We thus arrive at a 'kit' of parts, as shown in Fig 7.3(a). It is easy to verify by a few trials that the only two ways in which the kit can be assembled are those shown in Figs. 7.3(b) and 7.3(c). By introducing first one double hinge and then two, we obtain from one of these assemblies the two chains having double joints (Figs. 7.3(d) and 7.3(e)). Again, it is easy to verify by trial that introducing double hinges in the other simple chain will only produce the same results. We have now found all the 6-bar chains with $M = 1$. We can then find all possible 6-bar mechanisms, as in Fig. 7.4, by fixing each link in turn; there are ten of them.

Before going further let us consider how we knew that third-order links were the highest that could be expected. The general rule is that in an n-link chain with $M = 1$ the highest order of any link is $n/2$. This can easily be understood by considering Fig. 7.5. Each closed loop must have at least four links, or it will be a triangle and hence immovable. Thus for a link of order two we must have $n = 4$ at least, and for each unit increase in the order of the frame link we must, as is obvious from the figure, increase n by two. Since M is not affected by the choice of frame the result is generally true.

We now return to the linkages in Fig. 7.4 and consider in how many ways they can be used to mount an implement on a tractor, remembering that in each case

Fig. 7.4. The four types of 6-bar chain in Fig. 7.3 provide ten mechanisms.

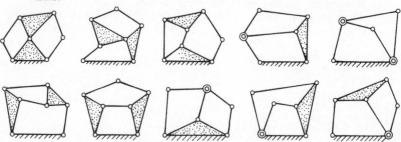

Fig. 7.5. The order of the highest-order link in a plane linkage with mobility $M = 1$ cannot be greater than half the total number of links.

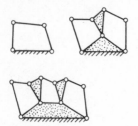

the fixed link will be the tractor itself, and that the implement must be mounted on a floating link, i.e. it must not be directly pivoted to the frame. This is done by considering each mechanism in turn, and one such case is shown in Fig. 7.6. By proceeding in this way we obtain, by a simple routine process, all the possibilities. They are not shown here, since the reader may like to try the method for himself; the full list of solutions is given by K. Hain, *Applied Kinematics*, London: McGraw-Hill, 1967, who also gives other examples of the use of systematics in obtaining basic designs.

By an extension of the basic principles explained in this chapter, it has been shown that there are 16 distinct 8-bar chains, giving by inversion 71 different mechanisms. There are 230 distinct 10-bar chains, but the total number of mechanisms that can be derived from them has not been published so far. The point brought out by this kind of analysis is that the number of alternatives open to the designer at the important stage of basic design is much larger than is usually realised, and certainly beyond anything that could be obtained by even the best of intuitive inventors. Mechanical engineering is not, as is often thought, an old and well-known art, but is still in its infancy since only a small fraction of the number of mechanisms revealed by systematics have ever been studied or used.

Fig. 7.6. Two alternatives to the design in Fig. 7.2 can be obtained from just one of the 6-bar linkages shown in Fig. 7.4.

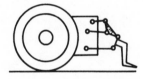 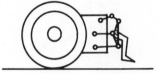

Exercises 7

7.1. Complete the analysis of the problem of mounting an implement on a tractor as discussed in this chapter. Do you consider any of the alternative arrangements preferable to the original layout shown in Fig. 7.2?

7.2. Fig. 7.7 shows the essential features of a pneumatic riveting press. Explain why this rather elaborate mechanism is used instead of direct application of the pneumatic cylinder to the riveting tool. Make a symbolic diagram of the mechanism and sketch all alternative arrangements, including those with multiple hinges, that could in principle do the same job. Would you, if designing such a press, prefer any of them to the original layout?

Fig. 7.7.

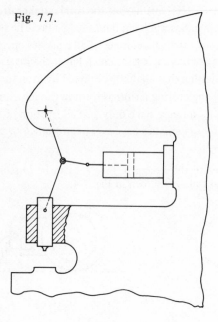

8 THE CHEBYSHEV-GRUBLER FORMULA

As stated in Chapter 1, all mechanisms are made up of components known as
links or bars, which contact each other in kinematic pairs. The number of links
and the number of pairs are connected by the Chebyshev-Grübler formula,
quoted in Chapter 1 without proof. A proof will now be given, together with an
account of certain exceptional cases in which the formula fails.

The positions of the various links in a mechanism relative to the frame are
determined when certain dimensions are specified, such as the four angles in the
linkage shown in Fig. 8.1. In this example only one of those angles can be specified independently, i.e. only one input can be applied to the mechanism, and thus

Fig. 8.1. The relative positions of the links in a 4-bar chain are determined if a single angle is known.

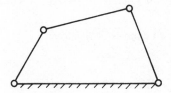

Fig. 8.2. The position of a moving link relative to the frame depends on
three measurements.

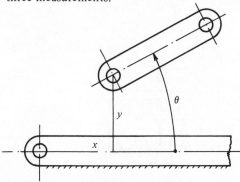

the mobility is given by $M=1$. Following up this idea, the mobility formula will be proved by considering the number of dimensions needed to specify the positions of the links in a mechanism before they are connected at the joints, and the way in which that number is reduced as the mechanism is assembled.

Consider first of all a frame and one other link, as in Fig. 8.2. To specify the position of the moving link relative to the frame we require three measurements. We can use, for example, the measurements shown in the figure: the two co-ordinates of a point in the moving link and the angle that a line in that link makes with the frame. Another moving link will require another three measurements, and so in general n links (of which one is taken as the frame) will require $3(n-1)$ measurements to determine their relative positions. All these measurements are independent, so the mobility of the system before any joints are connected up will be $3(n-1)$.

Now the effect of introducing a joint between two links is to reduce the mobility of one relative to the other. If, for example, the links are pivoted together, a single angle determines their relative position (Fig. 8.3(a)). Thus the mobility has been reduced from 3 to 1; the effect of the joint is to subtract 2 from the mobility. A similar effect is produced by any of the kinematic pairs shown in Fig. 8.3, for each of them connects the two links together in such a way that a single measurement determines relative position. These pairs are the revolute pair (pivot), (a), sliding pair (slider, or piston in a cylinder), (b), wrapping pair, (c) (the taut run of a belt moving without slip on a pulley, or the taut run of a chain

Fig. 8.3. Lower pairing removes two degrees of freedom.

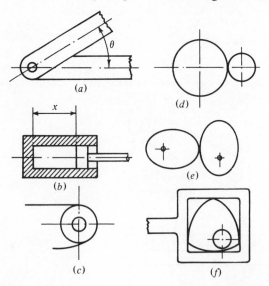

on a sprocket), the rolling pair (rolling without slip), (d) and (e), and a constant-diameter or lobed figure rotating inside a square hole, (f). All these can be called *lower pairs*, although some writers still follow an earlier usage of confining this term to pairs such as a plain bearing in which there is area contact. Since each lower pair reduces the mobility by 2, it follows that if there are j_l lower pairs the mobility will be reduced by $2j_l$.

The kinematic pairs shown in Fig. 8.4 are a point-contact or line-contact as in gear-teeth or a cam mechanism, (a), which allows combined rolling and slip; a constant-diameter figure working in a slot, (b), and a cord or belt kept taut but allowed to slip (c), for example, running over a stationary guide. These will be called *higher pairs*, a term originally confined to cases of line or point contact.

Fig. 8.4. Higher pairing removes only one degree of freedom.

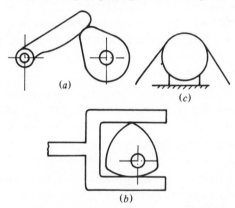

Fig. 8.5. The differential train has mobility $M = 2$; the other mechanisms shown here have $M = 1$.

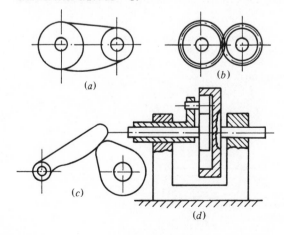

When two links are connected together by such a pair two measurements are needed to specify their relative position, and hence the pair removes only one degree of freedom. It follows that a total of j_h higher pairs will reduce the mobility simply by j_h.

Combining these three results we see that the mobility of a mechanism when all the joints are connected will be given by the Chebyshev–Grübler formula.

$$M = 3(n - 1) - j_h - 2j_l \tag{8.1}$$

Some examples of the use of this formula were given in Chapter 1, and further examples are given here (Fig. 8.5). The belt drive, (a), has four links (the frame, the two pulleys, and the taut run of belt) and four lower pairs (the two bearings and the two belt-pulley contacts), so $M = 3(4-1) - 2(4) = 1$. The gear-train, (b), has three links (the frame and two gears), two lower pairs (the bearings) and one higher pair (the contact between gear-teeth), so $M = 3(3-1)-1-2(2) = 1$. The same analysis applies to the cam mechanism, (c). In the epicyclic train used as a differential, (d), i.e. with all three shafts rotating, there are five links (the frame, three gears and the arm), two higher pairs (gear-tooth contacts) and four lower pairs (bearings), so $m = 3(5-1) - 2 - 2(4) = 2$.

There are, as previously stated, cases in which equation (8.1) fails to give a true value of M. These usually arise because, as a joint is connected up in the course of assembly, all that happens is that a point on a moving link is constrained to travel on a path it would take anyway. Hence that joint has no effect. The simplest and probably the most important example is shown in Fig. 8.6. In the upper left-hand diagram there is an assembly of five links and six pairs, which has $m = 0$ according to the formula and is in fact a rigid frame. But if the links are proportioned so that opposite links are parallel to each other, as in the upper right-hand diagram, the assembly is movable. This is because the bar PQ has no effect, the proportions of the other bars being such that the point Q would in

Fig. 8.6. Parallel linkages are useful exceptions to the Chebyshev–Grübler formula.

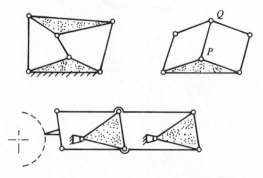

any event move in a circle of radius PQ about P. Any number of links can be added provided parallelism is maintained, and a number of useful mechanisms use this principle – locomotive coupling rods, movable trellises, etc. An interesting example is shown in the lower diagram. This mechanism is used to guide a cutting tool in a circular path about a 'virtual centre', for example a truing diamond for dressing a grinding wheel or a spherical turning attachment for a lathe. The formula would give $M = -1$.

There are two other important exceptions to the Chebyshev–Grübler formula:

(a) Two rotating disks (friction wheels), or two contour cams, can roll together without slip. There are three links (including the frame) and three lower pairs, so $M = 0$, but the device is movable because the point of contact is on the line of centres (the Aronhold–Kennedy theorem). If the point of contact is off the line of centres, as in a cam mechanism (Fig. 8.5), motion cannot occur without slip; the slip converts the contact into a higher pair and gives, correctly, $M = 1$.

(b) The mechanism in the left-hand diagram of Fig. 8.7 also has three links and three lower pairs, but is movable. This is because it can be derived from the mechanism shown in the right-hand diagram, which has a true value of $M = 1$, by bringing additional points on the slider into contact with the frame. Since these points are already moving parallel to the surface of the frame bringing them into contact will not have any effect.

Fig. 8.7. A wedge mechanism which is an exception to the Chebyshev–Grübler formula.

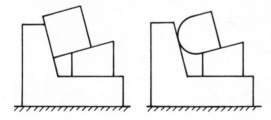

Exercises 8

8.1. Use the Chebyshev–Grübler formula to prove that the 6-bar linkage in Fig. 8.8 has $M = 1$. Verify this result by considering how you would draw the linkage, or assemble a model, for a given value of (*a*) the angle θ, and (*b*) the angle ϕ. By considering case (*a*) show that, in general, such a linkage can be assembled in four distinct ways (called the *closures* of the linkage, because the connections make it into a closed chain). Can you prove the same result by considering case (*b*)? Are there any special cases in which the linkage could have 3, 2, 1 or 0 closures? What is the practical significance of these special cases?

Fig. 8.8.

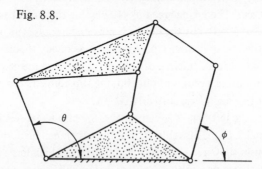

8.2. How would you draw the 8-bar linkage in Fig. 8.9 for a given value of θ? Explain why your method of drawing the linkage confirms that $M=1$ (as given by the Chebyshev–Grübler formula). How many closures has this linkage?

Fig. 8.9.

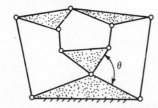

8.3. In the study of vehicle roll it is usual to make a first approximation by representing the front suspension as in Fig. 8.10. Each wheel, with its tyre and hub, is regarded as a rigid link (links 2 and 8) pivoting on the road at fixed points *A* and *B*. The instantaneous centre of link 5 (the vehicle body) relative to the road is called the *roll centre*. Show how the roll centre can be located provided the system can be drawn for a given roll angle θ. How would you make such a drawing? Calculate the mobility of the system and show that your proposed method of drawing confirms the calculated value.

Fig. 8.10.

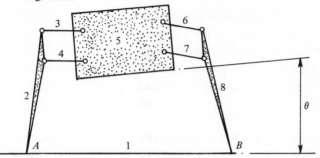

9 SYSTEMATICS OF MECHANISM

In Chapter 7 we showed how a whole class of linkages (for example, the 6-bar linkages) could be systematically enumerated, and the results used in the solution of a design problem. Mobility theory, which is the foundation of this work, was treated briefly in Chapter 1 and in greater detail in Chapter 8. With these foundations we can now consider how not only linkages, but also all other mechanisms, can be synthesised.

9.1 Deriving other mechanisms from linkages

Linkages are important not only for their own sake, but also because all other mechanisms can be systematically derived from them. This is done by changing lower pairs into higher pairs, by adding links, and by removing links, while keeping the mobility M and the number of closed loops in the kinematic chain unchanged. The following three simple rules are all that are needed:

1. *Any other lower pair can be substituted for a pivot*
Obviously this does not alter the mobility, the number of links, or the number of closed loops in the kinematic chain. The derivation of the slider–crank mechanism from the 4-bar linkage (Chapter 1) is an example of this rule.

2. *We can add a higher pair provided we remove one link and two pivots*
The proof is as follows. The mobility of the original linkage is given by

$$M = 3(n - 1) - 2j_l \qquad (9.1)$$

Now the modified linkage will have $(n - 1)$ links, $(j_1 - 2)$ lower pairs and one higher pair. Hence its mobility M' will be given by

$$M' = 3(n - 2) - 2(j_l - 2) - 1$$
$$= 3(n - 1) - 2j_l$$
$$= M$$

and thus the mobility is unaltered. In this way, for example, the simple cam mechanism is derived from the 4-bar linkage.

3. *We can add a link, provided we add one higher pair and one lower pair*
The proof follows the same lines as that for Rule 2, and is equally simple. An example of a mechanism obtained by the use of this rule will be given later in this chapter.

It can easily be proved that these three rules can be applied repeatedly. Their use will now be illustrated by deriving from the 4-bar linkage some other single-loop mechanisms with $M = 1$, which are the commonest and most useful types.

9.2   Single-loop mechanisms
As stated above, these will be derived from the 4-bar linkage. The linkage itself, however, can usefully be classified into a number of types and these will first be considered.

(a)   The 4-bar linkage
The theory of this classification, which depends on the relations between the lengths of the links, is given by various writers; only the main facts will be stated here. The most general type of 4-bar has all its links of different lengths. Using the notation of Fig. 9.1(*a*), they are labelled so that $a>c>d>b$. The most useful condition is when the links conform to *Grashof's criterion*, $(a + b)<(c + d)$. This is most easily remembered in words as '(longest link plus shortest link) is less than

Fig. 9.1. The Grashof chain and its inversions.

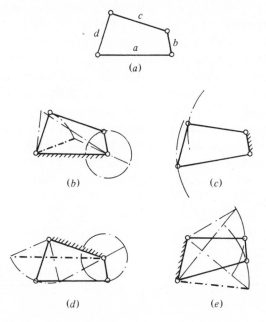

(sum of other two links)'. The behaviour of the mechanism depends on which link is fixed. With *a* fixed (Fig. 9.1(*b*)) link *b* is a crank and link *d* a rocker. With *b* fixed (Fig. 9.1(*c*)) both *a* and *c* are cranks; the linkage is called a *double-crank mechanism*, or *drag-link*. With *c* fixed (Fig. 9.1(*d*)) we obtain another crank-rocker mechanism, again with *b* as crank and *d* as rocker. Finally, with *d* fixed we obtain a *double-rocker mechanism* (sometimes called a *double-lever* mechanism) in which both *a* and *c* can only oscillate through limited angles (Fig. 9.1(*e*)).

When a 4-bar has been designed by any of the methods given in previous chapters a check should be made to see that the link-lengths do in fact allow the linkage to function acceptably throughout the cycle - say as a crank-rocker or double-crank if continuous rotation is required. Our design methods only ensure that the links will be in specified positions at specified points in the cycle; they give no assurance that they can move continuously between those points.

Apart from the parallel-crank linkage (mostly used for coupling locomotive wheels) the inversions of Grashof's chain provide the linkages most commonly required.

We now proceed to derive other single-loop mechanisms by applying our three rules to the 4-bar linkage.

(*b*)   *Mechanisms derived by Rule 1*

Replacing a pivot by a slide, we obtain the slider-crank chain (Fig. 9.2). In most applications the pivot is arranged to lie on the centre-line of the slideways, as in the lower diagram. The inversions of this modified slider-crank chain were described in Chapter 1, and those of the more general type are very similar.

A further application of Rule 1 enables us to introduce two sliders. This provides two different chains (Fig. 9.3(*a*) and (*b*)) depending on whether the order of the kinematic pairs is revolute-slider-slider-revolute, or revolute-slider-revolute-slider. Using an established notation, we shall refer to these as the RSSR and RSRS chains respectively. The RSSR chain has an important application as *Oldham's coupling* (Fig. 9.3(*c*)), which provides constant-speed transmission between parallel but misaligned shafts. At first sight the connection between this mechanism and the symbolic diagram may be puzzling. But if the coupling were to be viewed along the axis of one of the shafts it could be seen that it is a plane

Fig. 9.2. Slider-crank chains.

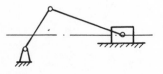
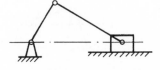

mechanism with four links (including the frame), and that the four kinematic pairs occur in the correct order. A practical form of the RSRS mechanism is obtained by shortening two of the links as in Fig. 9.3(*d*). This mechanism, known as *Rapson's slide*, is used as a marine steering-gear. The slider in contact with the frame is pulled along by chains attached to the steering-engine and the motion is transmitted via the other slider to the tiller. By considering the equilibrium in any one position, and applying the principle of virtual work, it can be seen that the torque applied to the tiller increases as the rudder is rotated from the centre position. The maximum steering effort is thus applied when it is needed.

Since the contact between a pulley and a taut run of belt is a lower pair, we can use Rule 1 to obtain the *band mechanisms* shown in Fig. 9.4. The ordinary belt drive using circular pulleys is of course the commonest type. Applications of the others are rare.

Rolling pairs can be introduced to give the two types of cam mechanism (with followers assumed to rotate without slip) shown in Fig. 9.5. As shown in the same figure, a cam mechanism can also be obtained by the use of a constant-diameter (lobed) figure rotating in a square hole, which gives a positive drive for which no spring is needed. Finally, by introducing two rolling pairs (Fig. 9.6 (*a*)) we obtain a simple vehicle; by introducing three (Fig. 9.6(*b*)), a device which

Fig. 9.3. The RSSR chain (*a*) giving Oldham's coupling (*c*) and the RSRS chain (*b*) giving Rapson's slide (*d*).

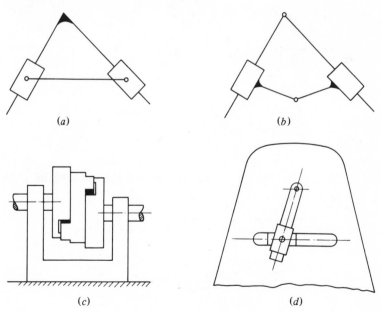

does not seem to be used; and the use of four rolling pairs gives us one of the earliest mechanisms, a means of moving heavy objects by the help of rollers (Fig. 9.6(*c*)).

(c)   *Mechanisms derived by Rule 2*
Replacing a pivot by a roll-and-slip contact, and removing a link and one other pivot, we obtain a cam mechanism (Fig. 9.7(*a*)) and a simple gear-train (Fig. 9.7(*b*)). Applying the same transformation to a slider–crank mechanism we obtain

Fig. 9.4. Band mechanisms.

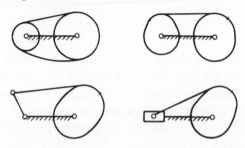

Fig. 9.5. Mechanisms using one rolling pair or a lobed figure in a square hole. Each has four links and four lower pairs.

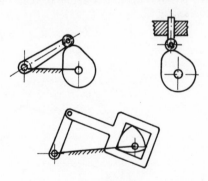

Fig. 9.6. Mechanisms using two, three and four rolling pairs.

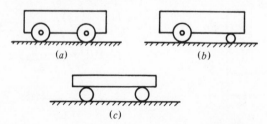

Fig. 9.7. Three-link mechanisms, each with two lower pairs and one higher pair.

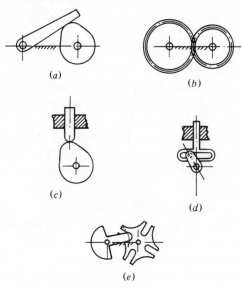

Fig. 9.8. The simplest possible mechanisms. Each has two links and two higher pairs.

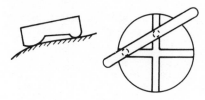

Fig. 9.9. Primitive mechanisms using a taut rope slipping over a guide.

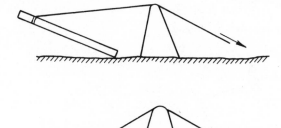

another cam mechanism (Fig. 9.7(*c*)), the scotch yoke mechanism (Fig. 9.7(*d*)) and the Geneva mechanism (Fig. 9.7(*e*)). We can obtain additional mechanisms by curving the slot in the scotch yoke (giving a very useful mechanism found in sewing-machines as a needle-drive), and by replacing the cam-and-follower pair with a lobed figure working in a slot to give positive drive. The use of these lobed figures is uncommon.

Using two roll-and-slip contacts (Fig. 9.8) we obtain a very simple 2-bar mechanism, found in nature when a rough object such as a stick is dragged over the ground, or in man-made form as the elliptic trammel.

Using a taut run of cord allowed to slip over a stationary guide (Fig. 9.9) we obtain a primitive method of raising a column (using two 'pivots'), or of dragging a sledge along the ground up to the foot of an obstacle (using one 'pivot' and a 'slider').

Some of the 'mechanisms' derived here belong to ergonomics rather than machine design. The sketch (Fig. 9.10) shows how a simple treadle-operated grindstone together with its operator forms a 6-bar linkage. By analysing a manually operated device in this way, which takes account of the fact that the operator is part of the kinematic chain, it is sometimes possible to achieve a better understanding of the design and thus to improve it.

Fig. 9.10. Kinematic analysis of a manually operated machine. The operator's hip joint acts as a fixed pivot, and his knee and ankle joints as moving pivots.

Fig. 9.11. The geared 5-bar linkage.

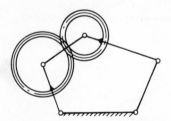

(d)   *Mechanisms derived from Rule 3*

This rule can be applied, if necessary several times in succession, to the original 4-bar linkage or to any mechanism derived from it by the other rules. Since the rule requires us to add links and pairs, and to take nothing away, an infinite number of mechanisms can be derived. Obviously only those of limited complexity will be of interest. The simplest example is the geared 5-bar chain shown in Fig. 9.11. Since the designer has at his disposal an extra link-length, a gear-ratio and the initial setting of the gears in engagement, the mechanism offers greater possibilities than the 4-bar. It is, however, much more difficult to design.

(e)   *Mechanisms which are exceptions to the Chebyshev-Grübler formula*

The additional single-loop mechanisms that occur as exceptions to the general mobility rule (friction wheels, contour cams and the wedge mechanism) have already been described in Chapter 8.

9.3   Applications of systematics

A simple example of the use of systematics to discover alternatives to a given design was described in Chapter 7. Perhaps the most striking of all uses was made by Wankel who systematically investigated possible rotary engine designs and produced the commercially viable result which had always eluded intuitive inventors.

The main use of systematics, however, lies not in solving particular design problems as they arise, but in inventing mechanisms in advance and classifying them in an orderly manner. This may seem to be an academic activity carried out for the amusement of those who do it, but in fact it is a very useful contribution to engineering practice. Firstly, inventing in advance of our requirements is useful because there is usually no time for inventing when a job has to be done. The important stage of basic design is often the most hurried; the designer is urged to commit himself so that his employer can tender for a job by a certain date. In these circumstances it is essential to have all the alternative approaches at one's finger-tips. We have, of course, in this introductory account, synthesised several mechanisms that have been known for centuries. But those mechanisms took centuries to evolve by intuitive invention, whereas now we can synthesise them rapidly and easily. And we can also synthesise many new types. Secondly, the number of usable mechanisms is so large that no designer can remember all of them, and it has become essential to prepare properly classified data so that information on mechanisms can easily be found. The old method, used in engineers' handbooks, of classifying mechanisms by use is unsatisfactory because each mechanism has a variety of uses and any proposed use can be met by a variety of mechanisms. A comprehensive catalogue of mechanisms which avoids duplications can only be based on classification by 'structure', i.e. the mobility and the number of loops in each mechanism. The old 'cookery-book' approach cannot enable us to take advantage of modern possibilities or face up to modern competition.

Exercises 9

9.1. Some of your solutions to the design problems in earlier chapters will be 4-bar linkages. Use Grashof's criterion to find if they are capable of continuous input-link rotation.

9.2. What equivalent for Grashof's criterion can be applied to the slider-crank mechanism? (Consider the slider–crank as a limiting case of the 4-bar when two links become infinitely long, thinking about what happens when they are very long but not yet infinite.)

9.3. How would you find the rubbing velocity in Oldham's coupling?

9.4. Derive all possible cam mechanisms from the 6-bar linkage and suggest any uses for them that you can imagine.

10 AN INTRODUCTION TO ANALYTICAL METHODS

In practical design work graphical methods are usually best; they are quick, and accurate enough for most purposes. There are, however, useful things that can only be done analytically. A very complex mechanism, for example the needle-bar drive mentioned in Chapter 1 (Fig. 1.2), may be beyond the scope of graphical analysis or synthesis.

There are several approaches to this subject. First of all, in the case of very simple mechanisms, we can use the ordinary methods of trigonometry. A slider–crank mechanism, for example, can be represented simply by a triangle and analysed by the sine and cosine rules. With more complex mechanisms attempts to use trigonometry by triangulating the figure become too cumbersome to be worthwhile, so we turn to more general methods. Of these, the vector-loop method, to be described in this chapter, was the first to be developed and is probably still the most commonly used. We shall illustrate it by an example: the 4-bar linkage.

10.1 Vector-loop equations

We shall treat the lines representing the links of a mechanism as vectors, measuring the components of each vector in rectangular co-ordinates attached to the frame. Thus, as in Fig. 10.1, if the link is of length x and makes an angle ϕ with the frame, its components are $x \cos \phi$ horizontally and $x \sin \phi$ vertically.

Fig. 10.1. Resolution of a vector into components; note the sense of the arrow-heads and the angle.

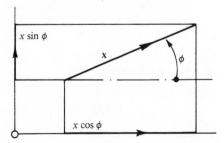

Note that we indicate a direction for the vector by an arrow-head on the link. If the arrow-head in our figure were pointing the other way the components would have negative numerical values. The angle ϕ is measured as the counter-clockwise angle *from* the frame *to* the link. If, at the end of a computation, we had the result $\phi = 210°$ it would mean that the link (and the arrow-head) was placed as in Fig. 10.2. The result $\phi = -150°$ would have exactly the same interpretation, since a negative numerical value of the angle means a clockwise rotation.

With these conventions clearly understood we label the 4-bar linkage as in Fig. 10.3. Now the loop of vectors is closed, so the sum of the vectors must be zero. Taking the senses of the arrow-heads into account we therefore have the vector equation

$$-\mathbf{a} + \mathbf{b} + \mathbf{c} - \mathbf{d} = 0 \qquad (10.1)$$

the signs being explained by the fact that we went round the loop clockwise. This is quite arbitrary, and of no importance; if we had chosen to go round counter-clockwise we would simply have changed signs right through the equation, which of course makes no difference.

There are two ways of interpreting this equation. We can split it into horizontal and vertical components, giving

$$-a + b \cos\theta + c \cos\psi - d \cos\phi = 0 \qquad (10.2)$$

$$b \sin\theta + c \sin\psi - d \sin\phi = 0 \qquad (10.3)$$

or express it in complex-number form, giving

$$-a + be^{i\theta} + ce^{i\psi} - de^{i\phi} = 0 \qquad (10.4)$$

Let us now consider some applications of these equations.

10.2 Analysis

Our purpose is to find a method of calculating θ when we are given ϕ, or ϕ when we are given θ; in other words, to find the relation between input and output.

Fig. 10.2. Interpretation of the direction angle of a vector; a negative angle must be measured in a clockwise sense.

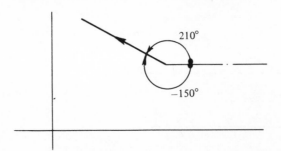

Having done this we can then, if we wish, find ψ by substitution in equation (10.3). We can, in fact, solve equations (10.2) and (10.3) to find the relation between any two of the angles θ, ψ and ϕ if we first eliminate one of them. To eliminate ψ we take the terms involving it over to the right-hand side in each equation, giving

$$- a + b \cos \theta - d \cos \phi = - c \cos \psi$$
$$b \sin \theta - d \sin \phi = - c \sin \psi$$

and then square and add these equations, giving

$$K_1 \cos \phi + K_2 \cos \theta + K_3 - \cos \theta \cos \phi - \sin \theta \sin \phi = 0 \qquad (10.5)$$

where $\quad K_1 = a/b$

$$K_2 = -(a/d)$$
$$K_3 = (a^2 + b^2 - c^2 + d^2)/2bd$$

To solve this trigonometric equation we transform it into an algebraic one by using the substitution

$$\sin \alpha = 2t/(1 + t^2)$$
$$\cos \alpha = (1 - t^2)/(1 + t^2)$$

where α is any angle and $t = \tan (\alpha/2)$. Putting $\tan (\theta/2) = u$ and $\tan (\phi/2) = v$ we get

$$(K_3 - K_2 - K_1 - 1)u^2v^2 + (K_3 - K_2 + K_1 + 1)u^2$$
$$+ (K_3 + K_2 - K_1 + 1)v^2 - 4uv + K_1 + K_2 + K_3 - 1 = 0 \qquad (10.6)$$

Now suppose θ is the given input angle. Then u is known and equation (10.6) is simply a quadratic in v, and solving it for v we can find ϕ. Similarly, if ϕ is the given angle we have a quadratic in u and can find θ. In either case there will be (in general) two solutions, since a quadratic has two roots. If these are real they correspond to the two ways in which the linkage can be assembled, as shown in Fig. 10.4 where we have taken θ as the input angle. The calculation is within the scope of any programmable calculator, let alone a computer, and can rapidly

Fig. 10.3. Lengths, angles and arrow-heads for the vector analysis of a 4-bar linkage.

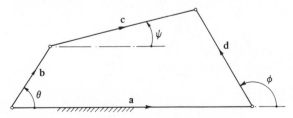

provide a table of values or a graph of corresponding angles throughout the cycle.

Mechanisms with more than one loop give sets of equations that cannot be solved so easily. These equations can, however, be solved by numerical methods.

We can obtain velocity equations by differentiating equations (10.2) and (10.3) with respect to time. Suppose θ is the input angle and $\dot\theta$ the input angular velocity, then we get

$$- c \sin \psi\, \dot\psi + d \sin \phi\, \dot\phi = b \sin \theta\, \dot\theta$$
$$c \cos \psi\, \dot\psi - d \cos \phi\, \dot\phi = - b \cos \theta\, \dot\theta \tag{10.7}$$

Now these are linear algebraic equations in the unknowns $\dot\phi$ and $\dot\psi$ and their solution is a matter of routine. A second differentiation gives equations for the angular accelerations. These, too, are linear algebraic equations in the unknown angular accelerations $\ddot\phi$ and $\ddot\psi$.

The result of differentiating the complex-number form of the vector-loop equation (10.4) is interesting. We get

$$i\dot\theta b e^{i\theta} + i\dot\psi c e^{i\psi} - i\dot\phi d e^{i\phi} = 0 \tag{10.8}$$

The effect of the differentiation is to multiply each vector by its angular velocity and rotate it through 90° in a counter-clockwise sense. But this is the way we obtain velocity diagrams, and in fact equation (10.8) is simply the vector-loop equation of the velocity diagram for the 4-bar linkage.

A second differentiation of the complex-number form of the vector-loop equation gives the equation of the acceleration diagram. This can help our understanding of acceleration diagrams and perhaps give some help in drawing them, particularly when the difficult question of Coriolis terms crops up. Consider, for example, the mechanism shown in Fig. 10.5. Let P_2 be a point that moves along link 2, being always coincident with a point P_3 fixed in link 3, as in the small

Fig. 10.4. The two closures of a 4-bar linkage. When the input angle θ is given the linkage can be drawn, or assembled, in two distinct ways, corresponding to the intersection of circular arcs struck from A and B_0.

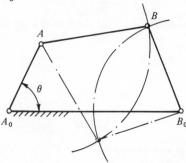

sketch. Both these points have *the same position at all times* so we have the equation

$$xe^{i\theta} = a + be^{i\phi} \qquad (10.9)$$

and differentiating twice (remembering that x is variable) we get

$$-(\dot{\theta})^2 xe^{i\theta} + i\ddot{\theta}xe^{i\theta} + \ddot{x}e^{i\theta} + 2i\dot{x}\dot{\theta}e^{i\theta} = -(\dot{\phi})^2 be^{i\phi} + i\ddot{\phi}be^{i\phi} \qquad (10.10)$$

Let us now pick out the significance of the various terms and see how they appear in the usual acceleration diagrams.

The first two terms on the left-hand side make up the acceleration of a point *fixed* on link 2, momentarily coincident with P_2. The next term is the sliding acceleration with which P_2 moves along link 2. The fourth term is the Coriolis

Fig. 10.5. Acceleration diagram involving Coriolis term and its relation to the vector-loop method.

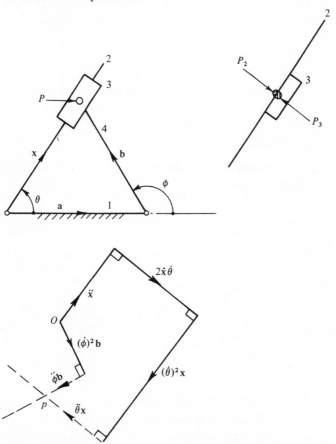

acceleration, obtained quite automatically without any knowledge of Coriolis' theorem being needed; moreover it appears with the correct sign when the numerical values of the terms are put in. The sum of all these four terms is the acceleration of P_2 relative to the frame. The two terms on the right-hand side are the acceleration of P_3, which is much easier to understand since this point simply moves in a circle of radius b.

The actual construction of the acceleration diagram depends on the fact that the velocity analysis has first been completed so that we know the numerical values (with signs) of $\dot{\theta}$, $\dot{\phi}$ and $\dot{x}$. This leaves, as unknowns, $\ddot{\theta}$, $\ddot{\phi}$ and $\ddot{x}$. Now a vector equation representing planar motion really amounts to two equations, because it can be split into components, so if any one of these accelerations or angular accelerations is given we can find the other two. Suppose, for example, we are given $\ddot{x}$, then we can draw in the four vectors shown in full lines in the acceleration diagram of Fig. 10.5. The remaining two, shown in broken lines, have unknown magnitudes but known directions and they intersect in the point P, which represents the head of the acceleration vector $\ddot{\mathbf{P}}$, to complete the diagram.

10.3 Synthesis

The equations that enable us to analyse a linkage can sometimes also be used for synthesis. We shall consider *Freudenstein's method*, applied to the 4-bar linkage as follows. We re-write equation (10.5) as

$$\cos \phi \, K_1 + \cos \theta \, K_2 + K_3 = \cos (\theta - \phi) \qquad (10.11)$$

which is a linear equation in the Ks. Now suppose we have to design a function generator for three corresponding pairs of input and output angles

input θ	output ϕ
θ_1	ϕ_1
θ_2	ϕ_2
θ_3	ϕ_3

we will have a set of three equations, namely

$$\begin{bmatrix} \cos \phi_1 & \cos \theta_1 & 1 \\ \cos \phi_2 & \cos \theta_2 & 1 \\ \cos \phi_3 & \cos \theta_3 & 1 \end{bmatrix} \begin{bmatrix} K_1 \\ K_2 \\ K_3 \end{bmatrix} = \begin{bmatrix} \cos (\theta_1 - \phi_1) \\ \cos (\theta_2 - \phi_2) \\ \cos (\theta_3 - \phi_3) \end{bmatrix} \qquad (10.12)$$

which we can solve for the Ks. We could now calculate the actual link-lengths a, b, c and d from the definitions of the Ks in Section 10.2, were it not for the fact that we have four of these unknown quantities but only three equations relating them to the Ks. But since we are dealing only with angles the *size* of the linkage is of no significance, only its shape. So we put $a = 1$ (i.e. we take the frame as the unit of length) and we then get the rest of the link-lengths in terms of this unit as

$$b = 1/K_1$$
$$d = -(1/K_2)$$
$$c = \sqrt{(1 + b^2 + d^2 - 2bd\,K_3)}$$

We must remember that the purpose of such a linkage is not really to correlate three pairs of *angles*, but two pairs of *rotations*, so θ_1 and ϕ_1 are arbitrary. For example, suppose the specified rotations are

input	output
30°	20°
30°	35°

and we take, as arbitrary but 'reasonable' starting angles, $\theta = 60°$ and $\phi = 50°$. Then the table of angles becomes

θ	ϕ
60°	50°
90°	70°
120°	105°

Using these values we find the linkage dimensions and then draw the linkage in several positions within its proposed working range to make sure that its least value of transmission angle is satisfactory and, indeed, that it can remain assembled throughout the range. If the result is unsatisfactory we try new starting angles. The method is reasonably quick if the calculations are programmed, as can easily be done for any computer or programmable calculator.

10.4 The offset slider–crank mechanism

When the slider–crank mechanism has an offset (Fig. 10.6) we can no longer represent it simply by a triangle; it can, however, easily be analysed and synthesised by the vector-loop method. With the notation in the figure the vector-loop equation, resolved into horizontal and vertical parts, is

$$a \cos \theta - x - b \cos \phi = 0 \tag{10.13}$$
$$a \sin \theta - c - b \sin \phi = 0 \tag{10.14}$$

We want to find the relation between the crank angle θ and the slider displacement x, so it is necessary to eliminate ϕ. Taking the terms involving ϕ over to the right-hand side in each equation, squaring the equations and adding them we get

$$x^2 - 2a \cos \theta\, x + c^2 - b^2 - 2ac \sin \theta = 0 \tag{10.15}$$

a quadratic in the unknown x if the crank angle θ is given. Having found x, and already knowing θ, we can now find ϕ from either of the first two equations. There are, of course, two values of x corresponding to the two closures (Fig. 10.7).

Freudenstein's method is easily adapted to the synthesis of this mechanism. We write equation (10.15) as

$$2 x \cos \theta \, K_1 + 2 \sin \theta \, K_2 + K_3 = x^2 \tag{10.16}$$

where $K_1 = a$

$K_2 = ac$

$K_3 = b^2 - c^2$

and for three pairs of corresponding values of θ and x, substituted into equation (10.15), we obtain a set of three linear equations in the three unknown Ks. The two link-lengths and the offset c are then obtained from the definitions of the Ks as

$a = K_1$

$c = K_2/a$

$b = \sqrt{(K_3 + c^2)}$

Fig. 10.6. Vector-loop diagram for a slider–crank mechanism with an offset.

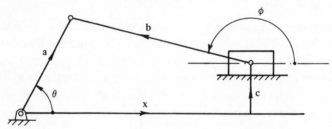

Fig. 10.7. The two closures of a slider–crank mechanism.

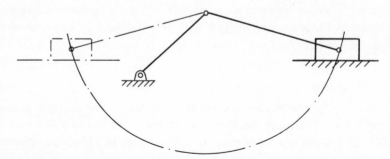

As in the case of the 4-bar function generator, the purpose of the mechanism is really to correlate two rotations with two displacements, so we can take arbitrary initial values for θ and x.

Exercises 10

10.1. Program equation (10.6) for a calculator or computer and check your program by drawing and measuring a 4-bar linkage. If computer graphic facilities are available write a program to plot a graph of output angle against input angle.

10.2. Use equations (10.7) to verify your graphical solution of Exercise 4.1(a).

10.3. Use Freudenstein's method to solve Exercise 2.2, taking an intermediate pair of shaft rotations based on the assumption that the mechanism is intended to replace spur gearing.

10.4. It is required to drive a slider through 10 mm while a crank rotates through 60° and through a further 10 mm while the crank continues to rotate through a further 45°. Use Freudenstein's method, modified as in equation (10.16), to design a slider–crank linkage for this purpose.

11 THE CONSTRAINT METHOD OF KINEMATIC ANALYSIS

The vector-loop method, described in Chapter 10, was the first method of kinematic analysis that offered a general alternative to graphical techniques. Its importance nowadays, however, lies mainly in its applications to synthesis and in the light it throws on velocity and acceleration diagrams. For the analysis of linkages, particularly those with more than four links, the recently introduced constraint method is usually simpler and therefore better.

The basic principle of the constraint method is that certain points on the links, such as bearing-centres, are constrained to move in particular paths because the links are rigid. These paths are often very simple – circles or straight lines – and can therefore be represented by simple algebraic equations. The method is closely analogous to the way in which we draw a linkage, locating pivot-centres by the intersection of circular arcs, and is therefore easy to understand and check. As in the case of the vector-loop method, we start with the simple example of a 4-bar linkage, but we shall see that the constraint method is easily extended to more complex linkages.

11.1 The 4-bar linkage

We label the linkage as in Fig. 11.1, taking the left-hand crank-centre as origin and the fixed link as the x-axis. The input angle θ is given and we wish to find the co-ordinates (x_1, y_1) and (x_2, y_2) of the coupler pivot-centres. The values of x_1 and y_1 are given immediately by

$$x_1 = b \cos \theta \tag{11.1}$$

$$y_1 = b \sin \theta$$

and we note that

$$x_1^2 + y_1^2 = b^2 \tag{11.2}$$

These equations, of course, express the fact that the point (x_1, y_1) must lie on a circle, centre the origin and radius b, and also on a straight line through the origin at an angle θ with the x-axis; which is exactly how we locate the point when drawing the linkage.

Now consider the way we would continue with our drawing to locate the other point (x_2, y_2). We strike an arc, centre (x_1, y_1) and radius c, representing the coupler, and an arc, centre $(a, 0)$ and radius d, representing the right-hand crank. In general, these arcs intersect in two points, giving the two possible locations for (x_2, y_2), i.e. the two closures of the linkage. Analytically this means that we must solve the equations

$$(x_2 - x_1)^2 + (y_2 - y_1)^2 = c^2 \tag{11.3}$$

$$(x_2 - a)^2 + y_2^2 = d^2 \tag{11.4}$$

Since x_1 and y_1 have already been calculated from equation (11.1) these last two equations have only two unknowns. By analogy with the drawing we expect them to have two solutions, which gives us a guide when doing the algebra; we expect them to reduce to a single quadratic equation.

Subtracting equation (11.4) from equation (11.3) we get

$$2(a - x_1)x_2 - 2y_1 y_2 = G \tag{11.5}$$

Fig. 11.1. Kinematic analysis of 4-bar linkage by the constraint method, with θ as input angle. The method is used to calculate the co-ordinates (x_1, y_1) and (x_2, y_2) of the coupler pivot-centres. Note the two possible closures: the open configuration is shown in solid lines and the crossed configuration in broken lines. The circular-arc construction used to draw the linkage for a given value of input angle corresponds to the solution of simultaneous quadratic equations in the constraint method.

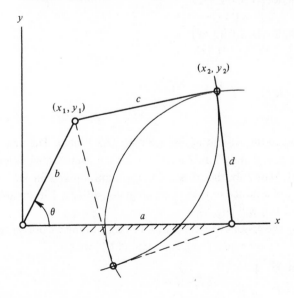

where

$$G = a^2 - b^2 + c^2 - d^2$$

and solving for x_2 we get

$$x_2 = (G + 2y_1 y_2) / 2(a - x_1) \tag{11.6}$$

Substituting this value of x_2 into equation (11.4) we get a quadratic in y_2. Finally, we solve this quadratic for y_2 and use equation (11.6) to calculate the corresponding value of x_2. The details are mere school algebra, but cumbersome, and the method is only worthwhile if we program it for a computer. Since we can use the calculation in a variety of ways (for example for a 4-bar linkage alone, for a 4-bar as part of a larger mechanism, for tabular or graphic output, etc) it is best to write the program as a subroutine, say as follows (using FORTRAN):

```
      SUBROUTINE FBAR (A,B,C,D,THET,I,X1,Y1,X2,Y2,J)
      X1=B*COS(THET)
      Y1=B*SIN(THET)
      G=(A+B)*(A-B)+(C+D)*(C-D)
      P=4*(Y1*Y1+(A-X1)*(A-X1))
      Q=4.*Y1*(G-2.*A*(A-X1))
      R=4.*(A-X1)*((A-X1)*(A+D)*(A-D)-A*G)+G*G
      DEL=Q*Q-4.*P*R
      IF(DEL.LT.O.) GO TO 4
      J=1
      GO TO (1,2)I
    1 Y2=(-Q+SQRT(DEL))/(2.*P)
      GO TO 3
    2 Y2=(-Q-SQRT(DEL))/(2.*P)
    3 X2=(G+2.*Y1*Y2)/(2.*(A-X1))
      RETURN
    4 J=2
      RETURN
      END
```

The input parameters to the subroutine are *A,B,C,D, THET* and *I*. The first five correspond to our a, b, c, d and θ, and *I* is set at 1 for the open configuration (solid lines in Fig. 11.1) or 2 for the crossed configuration (broken lines). The remaining parameters are returned to the calling program, with $J = 1$ if the linkage will assemble or $J = 2$ if it will not.

By differentiating the various equations with respect to time we obtain equations for the velocity components. From equations (11.1) we get

$$\left. \begin{array}{l} \dot{x}_1 = -b \sin \theta \dot{\theta} \\[4pt] \dot{y}_1 = b \cos \theta \dot{\theta} \end{array} \right\} \tag{11.7}$$

where $\dot{\theta}$ is the input angular velocity. From equations (11.3) and (11.4) we get

$$\begin{bmatrix} (x_2 - x_1) & (y_2 - y_1) \\ (x_2 - a) & y_2 \end{bmatrix} \begin{bmatrix} \dot{x}_2 \\ \dot{y}_2 \end{bmatrix} = \begin{bmatrix} (x_2 - x_1)\dot{x}_1 + (y_2 - y_1)\dot{y}_1 \\ 0 \end{bmatrix} \quad (11.8)$$

These equations are linear in the unknowns $\dot{x}_2$ and $\dot{y}_2$. The square matrix on the left-hand side will also occur when we calculate acceleration components. It is therefore useful to compute its determinant separately, by the single FORTRAN statement

$$DEL = Y1 * (X2 - A) + Y2 * (A - X1)$$

If *DEL* = 0, incidentally, we know that the linkage is in a limit position for the left-hand crank. We now solve equations (11.8) by the usual rule for simultaneous algebraic equations, the necessary subroutine being

```
SUBROUTINE FBARV (A,DTHET,X1,Y1,X2,Y2,DEL,DX1,DY1,DX2,DY2)
DX1=-Y1*DTHET
DY1=X1*DTHET
P=(DX1*(X2-X1)+DY1*(Y2-Y1))/DEL
DX2=Y2*P
DY2=(A-X2)*P
RETURN
END
```

Where the first seven parameters are input (*DTHET* being our $\dot{\theta}$) and the last four are $\dot{x}_1, \dot{y}_1, \dot{x}_2$ and $\dot{y}_2$, the output.

Differentiating again, we get equations for the acceleration components. The subroutine for calculating them is

```
SUBROUTINE FBARA (A,DTHET,D2THET,X1,Y1,X2,Y2,DX1,
    DY1,DX2,DY2,DEL,
C D2X1, D2Y1,D2X2,D2Y2)
D2X1=-Y1*D2THET-DY1*DTHET
D2Y1=X1*D2THET+DX1*DTHET
R=DX2*DX2+DY2*DY2
P=D2X1*(X2-X1)+D2Y1*(Y2-Y1)-R-DX1*DX1-DY1*DY1+2.*
    (DX1*DX2+DY1*DY2)
D2X2=(P*Y2+R*(Y2-Y1))/DEL
D2Y2=(P*(A-X2)+R*(X1-X2))/DEL
RETURN
END
```

Where *D2THET* is $\ddot{\theta}$, the input angular velocity, and *D2X1* ... *D2Y2* are the output parameters $\ddot{x}_1 \ldots \ddot{y}_2$.

11.2   The slider–crank linkage

As shown in Fig. 11.2, we choose the axes so that the crank-centre is at the origin and the slider axis is parallel to the x-axis. With the notation in the figure equations (11.1) to (11.3) still apply; the remaining equation is simply

$$y_2 = h \tag{11.9}$$

expressing the fact that the point (x_2, y_2) is constrained to move on a straight line parallel with the x-axis and distant h above it. Substituting this equation into equation (11.3) gives

$$x_2^2 - 2x_1 x_2 + b^2 - c^2 + h\,(h - 2y_1) = 0 \tag{11.10}$$

which relates the crank angle θ (since $x_1 = b \cos \theta$ and $y_1 = b \sin \theta$) to the slider position x_2.

If the crank angle θ is given equation (11.10) is solved as a quadratic in x_2, giving two solutions since there are two possible closures. If the slider position x_2 is given and we wish to find the crank angle θ we re-write the equation as

$$A \cos \theta + B \sin \theta = C \tag{11.11}$$

where

$$A = 2bx_2$$

$$B = 2bh$$

$$C = b^2 - c^2 + h^2 + x_2^2$$

To solve this equation we put $B/A = \tan \alpha$. Hence

$$\cos \alpha \cos \theta + \sin \alpha \sin \theta = \frac{C}{A} \cos \alpha$$

and therefore

$$\cos (\theta - \alpha) = \frac{C}{A} \cos \alpha \tag{11.12}$$

Fig. 11.2. Kinematic analysis of slider–crank mechanism by the constraint method. The input can be either the crank angle θ or the piston position x_2.

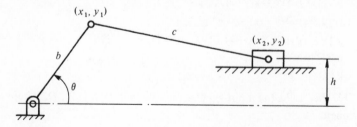

There are two solutions (since there are two values of α which satisfy the substitution $B/A = \tan \alpha$) and these correspond to the two closures.

Differentiation of equation (11.10) gives a relation between $\dot{x}_2$ and $\dot{\theta}$. A further differentiation gives a relation between $\ddot{x}_2$ and $\ddot{\theta}$.

11.3   Angles and coupler-points

The constraint method gives us the location of points on the links, but we often need to find the angle a link makes with the fixed link (Fig. 11.3). Obviously $\tan \phi = (y_2-y_1)/(x_2-x_1)$, but this alone is not sufficient to give us the value of ϕ, since there are two values of ϕ corresponding to a given value of $\tan \phi$. To decide which value to take we must consider the signs of numerator and denominator. Fortunately in most versions of FORTRAN there is a single statement that does the whole job, namely

$$\text{PHI} = \text{ATAN2}((Y2 - Y1),(X2 - X1))$$

We may also need to calculate the angular velocity and angular acceleration of the link, from the velocity and acceleration components of points on the link which are all that the constraint method provides directly. Suppose d is the distance between the two points on the link, then

$$\cos \phi = (x_2 - x_1)/d$$

and

$$\sin \phi = (y_2 - y_1)/d$$

By differentiating these equations with respect to time we get

$$\dot{\phi} = (\dot{x}_2 - \dot{x}_1)/(y_1 - y_2)$$

and also

$$\dot{\phi} = (\dot{y}_2 - \dot{y}_1)/(x_2 - x_1)$$

Fig. 11.3. Calculation of angles in the constraint method.

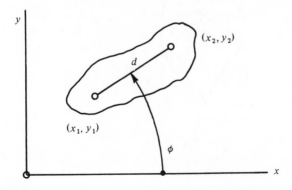

Both formulae are correct, but the first fails when $\phi = 0$ or $180°$ and the second when $\phi = 90°$ or $270°$. A FORTRAN function which makes the proper choice of formula is

```
FUNCTION ANGV(X1,Y1,X2,Y2,DX1,DY1,DX2,DY2)
IF(ABS((X2-X1)/(Y2-Y1)).LT.0.1)GO TO 1
ANGV=(DY2-DY1)/(X2-X1)
RETURN
1 ANGV=(DX1-DX2)/(Y2-Y1)
RETURN
END
```

By a second differentiation we obtain formulae for the angular acceleration $\ddot\phi$ in terms of the acceleration components of the points on the link, a suitable FORTRAN function being

```
FUNCTION ANGA(X1,Y1,X2,Y2,DX1,DY1,DX2,DY2,D2X1,D2Y1,
    D2X2,D2Y2)
IF(ABS((X2-X1)/(Y2-Y1)).LT.0.1) GO TO 1
ANGA=((X2-X1)*(D2Y2-D2Y1)-(DY2-DY1)*(DX2-DX1))/
    ((X2-X1)*(X2-X1))
RETURN
1 ANGA=((Y2-Y1)*(D2X1-D2X2)-(DX1-DX2)*(DY2-DY1))/
    ((Y2-Y1)*(Y2-Y1))
RETURN
END
```

Fig. 11.4. Calculation of the position, velocity and acceleration of an offset point on a link.

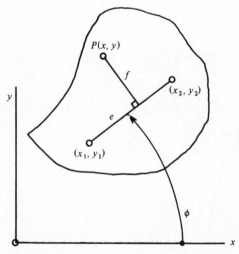

Next we consider the calculation of the position, velocity and acceleration of an offset point on a link (Fig. 11.4). Let (x_1, y_1) and (x_2, y_2) be the known co-ordinates of two points on the link and c be the distance between them; and let (e, f) be the local co-ordinates of the point P on the link, i.e. its co-ordinates in the moving link. We wish to calculate the co-ordinates (x, y) of P in the fixed axes on the frame. We start from the well known formulae in co-ordinate geometry

$$x = x_1 + e \cos \phi - f \sin \phi$$
$$y = y_1 + e \sin \phi + f \cos \phi \tag{11.13}$$

where ϕ is the angle between the axes, i.e. the rotation angle of the moving link. But $\cos \phi = (x_2 - x_1)/c$ and $\sin \phi = (y_2 - y_1)/c$. Substituting these values into our equations and translating into FORTRAN we have

```
SUBROUTINE POINT(X1,Y1,X2,Y2,C,E,F,X,Y)
X=X1+(E*(X2-X1)-F*(Y2-Y1))/C
Y=Y1+(E*(Y2-Y1)+F*(X2-X1))/C
RETURN
END
```

where the first seven parameters are input to the subroutine and the last two are output.

By differentiating equations (11.13) with respect to time we get equations for the velocity components of P in terms of the velocity components of the two base points on the link, and hence

```
SUBROUTINE POINTV (DX1,DY1,DX2,DY2,C,E,F,DX,DY)
DX=DX1+(E*(DX2-DX1)-F*(DY2-DY1))/C
DY=DY1+(E*(DY2-DY1)+F*(DX2-DX1))/C
RETURN
END
```

and by a further differentiation we get acceleration equations and hence

```
SUBROUTINE POINTA(D2X1,D2Y1,D2X2,D2Y2,C,E,F,D2X,D2Y)
D2X=D2X1+(E*(D2X2-D2X1)-F*(D2Y2-D2Y1))/C
D2Y=D2Y1+(E*(D2Y2-D2Y1)+F*(D2X2-D2X1))/C
RETURN
END
```

The above subroutines and functions can be combined in an endless variety of ways to suit particular problems. If one is doing a lot of work with linkages it is convenient to store the routines as a library on magnetic disc, rather than to put them in as cards with every job. It saves computer time and avoids the risk of cards getting shuffled, dog-eared or lost. Programming itself is then reduced to writing input, output and control statements, and calls to the routines. Doing this is rather like writing out a set of instructions for graphical analysis and intermediate calculations. By way of an example, suppose we wish to tabulate the path of a coupler point, and the right-hand crank angle and angular velocity, for the 4-bar linkage shown in Fig. 11.5. The input data will be the dimensions shown in the figure, $I = 1$ for the open configuration or $I = 2$ for the crossed configuration, constant increments of crank angle *THET* (in degrees) and constant angular velocity *DTHET*. A suitable program would be

```
      MASTER FOURBAR
      READ(1,100)A,B,C,D,E,F,I,THET,DTHET
      WRITE(2,200)A,B,C,D,E,F,I,THET,DTHET
      THETA=0
    1 THETB=THETA*3.14159/180.
      CALL FBAR(A,B,C,D,THETB,I,X1,Y1,X2,Y2,J)
      IF(J.EQ.2)GO TO 2
      DEL=Y1*(X2-A)+Y2*(A-X1)
      IF(DEL.EQ.O.)GO TO 2
      CALL FBARV(A,DTHET,X1,Y1,X2,Y2,DEL,DX1,DY1,DX2,DY2)
      CALL POINT(X1,Y1,X2,Y2,C,E,F,X,Y)
      PHI=ANG(A,O.,X2,Y2)
      PHI=PHI*180/3.14159
      DPHI=ANGV(A,O.,X2,Y2,O.,O.,DX2,DY2)
      WRITE(2,300)THETA,X,Y,PHI,DPHI
      THETA=THETA+THET
      IF(THETA.GT.360.)GO TO 2
      GO TO 1
```

Fig. 11.5. The subroutines and functions of the constraint method can easily be combined into programs for special purposes, such as the plotting of the coupler point position, the tabulation of the output angle ϕ and the tabulation of $\dot{\phi}$ in this linkage.

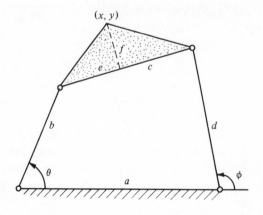

2 STOP
100 FORMAT(6FO.O,IO,2FO.O)
200 FORMAT(1H,6(E11.4,2X),IO,2(E11.4,2X))
300 FORMAT(1H ,5(E11.4,2X))
     END

It is surprising how many linkage analysis problems can be solved by combinations of the few simple subroutines and functions described above. There are, of course, not only other mathematical approaches to linkage analysis, but also other approaches to programming. At one extreme there is the practice of writing a complete individual program for each problem that arises, which is a reasonable thing to do if one has to deal with linkages very seldom. But if several programs are written in this way it soon becomes easy to see that much of the work is repetitive. At the other extreme there are large general-purpose programs which can only be implemented on very large computers. These programs are only worthwhile if a great deal of work on complex linkages is required.

11.4   Complex linkages
The constraint method can easily be extended to complex linkages, although a special computer program may have to be written to do the calculations. As an example consider the geared 5-bar linkage driving a slider (Fig. 11.6). Suppose the right-hand crank angle θ is the input. Then the left-hand crank angle ϕ is given by

$$\phi = n\theta + k$$

Fig. 11.6. The constraint method can be used for the analysis of complex mechanisms such as this geared linkage.

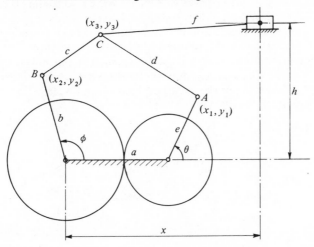

where n is the gear-ratio (having a negative numerical value in the case shown) and k is a constant depending on which teeth are engaged on assembly. We can calculate the co-ordinates of points A and B by the formulae

$$x_1 = a + e \cos \theta$$

$$y_1 = e \sin \theta$$

$$x_2 = b \cos \phi$$

$$y_2 = b \sin \phi$$

The co-ordinates of point C are now calculated by solving the pair of quadratic equations

$$(x_3 - x_1)^2 + (y_3 - y_1)^2 = d^2$$

$$(x_3 - x_2)^2 + (y_3 - y_2)^2 = c^2$$

These are solved for x_3 and y_3 by a similar procedure to that used for solving equations (11.3) and (11.4). The solution is completed by calculating x, the slider position, from the single quadratic equation

$$(x - x_3)^2 + (h - y_3)^2 = f^2$$

Exercises 11

11.1. Write a program for finding positions, velocities and accelerations of the mechanism shown in Fig. 10.5. Test your program by graphical analysis for particular values.

11.2. A connected pair of binary links (Fig. 11.7) is called a *dyad* and is an important sub-chain in most linkages. For example (right-hand diagram), a 4-bar linkage is a dyad driven by a crank. Write a subroutine DYAD (A,B,X1,Y1,X2,Y2,I,X,Y,J) to calculate X and Y when $A \ldots Y2$ are given; the integer value I is to indicate which of the two possible closures is to be taken and J is returned to the calling program as $J = 1$ if the subroutine works or $J = 2$ if the dyad cannot be assembled. Test your subroutine by writing a program for the 4-bar linkage. Note that DYAD allows us to put the fixed pivots anywhere we please, whereas the subroutines described in this chapter do not.

Fig. 11.7.

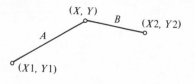
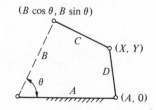

11.3. Extend the work of Exercise 11.2 to velocities and accelerations.

11.4. Write a program to calculate link positions in a 4-bar linkage for a given coupler angle (Fig 11.8) (Hint: consider the broken lines in the figure.)

Fig. 11.8.

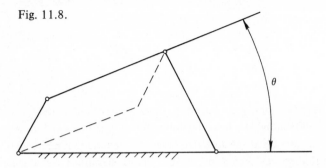

12 FORCES IN LINKAGES

Once we have designed a linkage it becomes necessary to calculate the forces and torques acting on the individual links, including the driving force or torque required to drive the linkage at the specified speed. We assume that a kinematic analysis of the linkage has been done, so that the accelerations and angular accelerations of the links are known. It is also necessary to have at least a rough design of the individual links so that we know their masses and moments of inertia.

Graphical methods of force analysis are described in books on the theory of machines. These methods, until recently the only ones available, are laborious, particularly if we have to repeat them for several positions of the input link so as to study the behaviour of the mechanism throughout its cycle. We shall limit ourselves here to the alternative approach, analytical methods which can be implemented on a computer or programmable calculator.

12.1 Mechanical principles

Whichever method is to be used, whether numerical or graphical, we depend on the following basic mechanical principles:

(*a*) The acceleration of the mass centre of a link of mass m, multiplied by the mass of the link, is equal to the resultant of the applied forces acting on the link. In vector notation this is the familiar equation $m\,\mathbf{a} = \mathbf{F}$. For calculations we need to separate this equation into components along the x- and y-axes, giving

$$m\,a_x = F_x \tag{12.1}$$

$$m\,a_y = F_y \tag{12.2}$$

(*b*) The angular acceleration of a link, multiplied by its moment of inertia about a perpendicular axis through its mass centre, is equal to the torque acting on the link. In symbols this is

$$I\alpha = T \tag{12.3}$$

Rules (*a*) and (*b*) are illustrated in Fig. 12.1. We also have (*c*) Newton's third law: action and reaction are equal and opposite.

(*d*) *If friction is negligible* the net power flow into a mechanism is equal to the rate of increase of the kinetic energy of the moving links plus the rate of increase of the potential energy. The potential energy, however, is usually ignored since it arises from the weights of the moving links, which are negligible in comparison with the inertial forces in a high-speed linkage. The formulae by which this rule is applied are considered below, in the section on energy methods.

12.2   Force analysis

The method can be illustrated by the following example of a slider–crank mechanism used as a pump. Fig. 12.2 shows the mechanism with the relevant dimensions, the given gas pressure P which acts as load and the required driving torque T which, together with the bearing loads, has to be calculated. We take the crankshaft centre as origin of co-ordinates, the x-axis being along the slider axis (cylinder centre-line). We assume that the crank is driven at a constant angular velocity ω and thus that we can calculate the acceleration components $(\ddot{x}_1, \ddot{y}_1)$, $(\ddot{x}_3, \ddot{y}_3)$ and $\ddot{x}_4$ of the mass centres of crank, connecting-rod and piston respectively, and the angular acceleration $\ddot{\phi}$ of the connecting-rod. These calculations are purely kinematic and can be done by any convenient method, such as that

Fig. 12.1. Force and torque acting on a link, and the acceleration and angular acceleration caused by them.

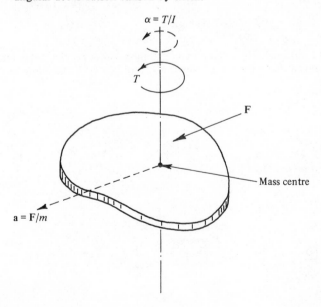

$$\alpha = T/I$$

T

F

Mass centre

$$a = F/m$$

described in Chapter 11. Let us assume that this essential first step has been completed.

We now sketch free-body diagrams for the links, showing the forces (Fig. 12.3). We can begin with any link. Let us take the crank first, and show the unknown forces X_1, Y_1, X_2 and Y_2, and the unknown driving torque T, acting on it. We now move to the connecting-rod, using Newton's third law to give us the forces acting on the big end; they are of magnitude X_2 and Y_2 acting in the opposite directions to those forces acting on the crank. There will also be two further unknown forces, X_3 and Y_3, acting on the little end. We have now arrived at the piston. Again we use Newton's third law to give us forces X_3 and Y_3 acting in opposite directions to those forces acting on the connecting-rod, and we also have the reaction Q of the cylinder wall and the given gas force P. Finally we have the forces on the frame, and also a torque, all given by Newton's

Fig. 12.2. Slider–crank mechanism operating as a pump. The crank centre O is the origin of rectangular co-ordinates used in kinematic and force analysis.

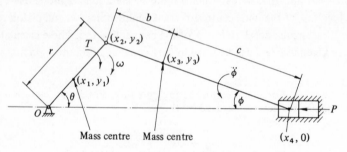

Fig. 12.3. Free-body diagrams and assumed forces for the slider–crank mechanism. Note the use of Newton's third law in the labelling of the forces.

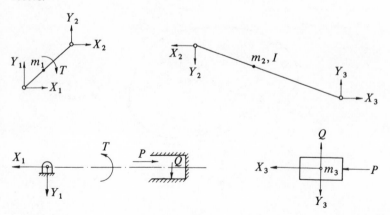

third law. Let us now count the unknown quantities. These are $X_1, Y_1, X_2, Y_2,$ X_3, Y_3, T and Q; eight of them, so we need eight equations.

In labelling the free-body diagrams we have *assumed* directions for the forces and for the unknown torque T, as indicated by the arrow-heads. These assumptions are quite arbitrary. If, when the calculations are finished, the numerical value of any force or torque is found to be positive, it means that our assumption concerning that particular force or torque was correct. If the numerical value is negative the direction must be reversed. Correct results will be achieved whatever directions we assume, provided we conform to Newton's third law as we move from link to link in our labelling of the free-body diagrams. The free-body diagram method is the key to success in this work; attempts to show all the forces on a sketch of the assembled mechanism are the usual cause of difficulty, confusion and error.

Having labelled the free-body diagrams we now apply Rules (*a*) and (*b*), in that order, to the crank, connecting-rod and piston. This gives us the following eight equations:

$$X_1 + X_2 = m_1\ddot{x}_1 \tag{12.4}$$

$$Y_1 + Y_2 = m_1\ddot{y}_1 \tag{12.5}$$

$$r\sin\theta\, X_2 - r\cos\theta\, Y_2 + T = 0 \tag{12.6}$$

$$-X_2 + X_3 = m_2\ddot{x}_3 \tag{12.7}$$

$$-Y_2 + Y_3 = m_2\ddot{y}_3 \tag{12.8}$$

$$b\sin\phi\, X_2 + b\cos\phi\, Y_2 + c\sin\phi\, X_3 + c\cos\phi\, Y_3 = I\,\ddot{\phi} \tag{12.9}$$

$$-X_3 = m_3\ddot{x}_4 + P \tag{12.10}$$

$$Y_3 + Q = 0 \tag{12.11}$$

These are linear algebraic equations in the eight unknowns, the right-hand sides being known quantities. The complete set can, of course, easily be solved using standard software on a computer. But it is always worthwhile to look at a set of equations to see if it breaks up into smaller, independent sets. In this case we see that (12.10) gives X_3 directly, and we can then get X_2 directly from (12.7). Using these values, (12.8) and (12.9) reduce to a set of two simultaneous equations giving us Y_2 and Y_3. The rest of the unknowns are now obtained directly, each from a single equation, (12.4), (12.5), (12.6) and (12.11) giving us X_1, Y_1, T and Q respectively. Solving the problem in this way is better even if we use a large computer, and essential with a very small machine such as a programmable calculator. Indeed, by looking at the set of equations instead of plunging straight into a solution, we have reduced the problem to something we could solve with the simplest calculator or even a slide-rule.

The method can be used with any plane linkage, or indeed any plane mechanism. To summarise, we first attend to the kinematics. Then we show the forces on free-body diagrams, using Newton's third law. Then we write down the three equations of Rules (*a*) and (*b*) for each moving link; and we check to make sure we have as many equations as unknowns. Then we look at the set of equations to see if it breaks up into smaller sets. Finally we solve the equations.

Two important points arise in the interpretation of the results of a force analysis. First, the forces and torque acting on the *frame* are not necessarily, or usually, in equilibrium and will therefore give rise to acceleration and angular acceleration of the frame. If the frame is sufficiently massive in comparison with the moving links these accelerations will be negligible. This case arises, for example, when we have a small mechanism securely bolted onto the frame of a machine tool which itself is bolted down onto concrete foundations, the effective frame being now the whole earth. When the frame is not so massive, for example in the case of an engine on flexible mountings, the frame itself will move and our calculations will no longer be strictly correct, although usually sufficiently so for design purposes. Secondly, although all the forces act in parallel planes, they do not all act in the *same* plane and therefore they will give rise to couples tending to rotate the mechanism about some axis in the plane of the drawing, or parallel to that plane. This effect is usually unimportant, but can be troublesome if the 'thickness' of the mechanism is considerable.

12.3 Force analysis by energy methods

The method we have just considered is the easiest to understand and use, since it depends on basic mechanical principles applied in a set routine. It enables us to calculate all the forces and torques, but puts us to considerable labour in doing so. If we are interested in only a single force or torque the energy method provides an alternative. Suppose, for example, we are only concerned to find the input torque T for the pump. We can do this by means of a single equation. First, as is always necessary, we solve the kinematic problem. Then we apply Rule (*d*). The kinetic energy of a moving link is given by $\frac{1}{2}m(\dot{x}^2 + \dot{y}^2) + \frac{1}{2}I\dot{\theta}^2$, where m is the mass of the link, I its moment of inertia, x and y the co-ordinates of its mass centre and $\dot{\theta}$ its angular velocity. The rate of change of the kinetic energy will therefore be

$$dE/dt = m(\dot{x}\,\ddot{x} + \dot{y}\,\ddot{y}) + I\dot{\theta}\,\ddot{\theta} \tag{12.12}$$

In our example the net power flow into the system is $T\omega - P\dot{x}_4$. So, applying Rule (*d*), the input torque T is given by

$$T = [m_1(\dot{x}_1\ddot{x}_1 + \dot{y}_1\ddot{y}_1) + m_2(\dot{x}_3\ddot{x}_3 + \dot{y}_3\ddot{y}_3) + I\dot{\phi}\ddot{\phi} + \\ m_3\dot{x}_4\ddot{x}_4 + P\dot{x}_4]/\omega \tag{12.13}$$

By way of a check we notice that if the pump is running very slowly, so that accelerations are negligible, equation (12.13) reduces to

$$T = P\dot{x}_4/\omega \tag{12.14}$$

which is the ordinary 'energy equation' of elementary theory.

The energy method can also be used to find a constraint force; let us use it to find the force Q exerted by the cylinder wall on the piston. We imagine the piston to have a small displacement of unit velocity perpendicular to the wall (Fig. 12.4). This will produce a power flow into the mechanism of amount Q, since the unit velocity is in the same direction and sense as the force. We now find, by a velocity diagram or calculation, the angular velocities $\dot{\theta}'$, $\dot{\phi}'$, and the velocities $\dot{x}'_1, \dot{y}'_1, \dot{x}'_3$ and $\dot{y}'_3$ which this 'virtual' unit velocity would produce. We note that $\dot{x}_4$ does not come into the problem because the displacement is normal to the cylinder axis. Using the virtual velocities thus obtained *and the real accelerations corresponding to what is actually happening* we get the power equation

$$Q = T\dot{\theta}' + m_1(\dot{x}'_1\ddot{x}_1 + \dot{y}'_1\ddot{y}_1) + m_2(\dot{x}'_3\ddot{x}_3 + \dot{y}'_3\ddot{y}_3) + I\dot{\phi}'\ddot{\phi} \tag{12.15}$$

which immediately gives us the required force Q. Note that this formula requires that we know the value of T, which must be calculated first, say by equation (12.13). The gas force and the piston inertia do not appear in our formula, since the displacement is perpendicular to their lines of action; but they are accounted for in the calculation of T. The theory of the method is outside our scope but the principles involved (virtual work and D'Alembert's principle) are explained in books on mechanics.

To find the forces acting at a moving bearing by the energy method we use Newton's third law to split the mechanism into two separate systems. The calculation of the big-end loads in our pump will show the method (Fig. 12.5). We examine the system consisting of the connecting-rod and piston, calculating

Fig. 12.4. Energy method of finding the side-thrust exerted by the cylinder on the piston.

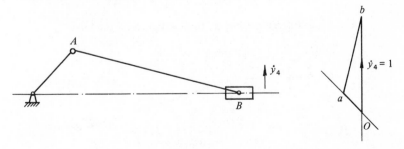

separately each of the force components X_2 and Y_2. The first of these, X_2, is
the easiest. We give the big end a horizontal displacement of velocity $\dot{x}'_2 = 1$.
Then the piston, and the mass centre of the connecting-rod, will also each have
unit virtual velocity, horizontally to the right; and the virtual angular velocity of
the connecting-rod will be zero, since the whole subsystem simply moves horizontally to the right like a single rigid link. The applied forces are X_2, which we
have to find, and P, which is given. Both these forces act against the direction of
motion so the work done by them on the subsystem is negative. The power
equation is therefore

$$X_2 = -(m_2\,\ddot{x}_3 + m_3\,\ddot{x}_4 - P) \tag{12.16}$$

We could, of course, have obtained this result simply by the direct application of
Newton's second law (Rule (a)) to the subsystem. But the calculation of the
other component, Y_2, is not so trivial and cannot be done directly by Rule (a).
In this case we give the big end a vertical displacement of velocity $\dot{y}'_2 = 1$,
causing virtual velocities $\dot{x}'_3, \dot{y}'_3$ to the mass centre of the connecting-rod and $\dot{x}'_4$
to the piston, and a virtual angular velocity $\dot{\phi}'$ to the connecting-rod. The power
equation is then

$$Y_2 = -(m_2\dot{x}'_3\,\ddot{x}_3 + m_2\dot{y}'_3\,\ddot{y}_3 + I\dot{\phi}'\ddot{\phi} + m_3\dot{x}'_4\,\ddot{x}_4 - P\dot{x}'_4) \tag{12.17}$$

As in the calculation of Q (and all applications of this method) we use the virtual
velocities but the *actual* accelerations.

   To calculate the big-end load by the energy method we used the subsystem
of connecting-rod and piston because it was possible to make the two displacements independently of each other. We could not use the other subsystem, consisting simply of the crank, because the only kinematically possible displacement
of the point of application of the force would be perpendicular to the crank
centre-line, which would not be in line with either force component. Then both
unknown force components would do work, and we would have two unknowns
but only one power equation. The essence of the method is to choose a way of
splitting the mechanism, and a way of applying a virtual displacement to one of
the subsystems, which will allow only one unknown force or torque to do work.

Fig. 12.5. Energy method of finding the crank-pin load.

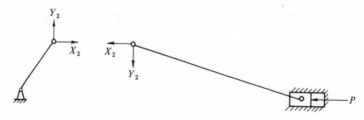

Exercises 12

12.1. Forces and a torque act as shown in Fig. 12.6 on a uniform rectangular link with a hole in it, the mass of the link being 30 kg. Calculate the angular acceleration of the link, and the acceleration of its mass centre.

Fig. 12.6.

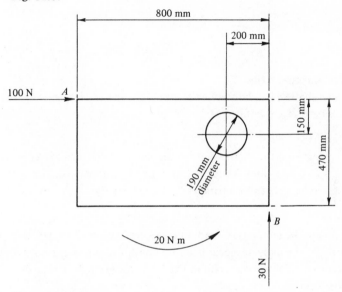

12.2. A link with the dimensions and mass of question 12.1 lies on a frictionless horizontal surface. A single force acts on the link, the line of action of this force being along the diagonal AB. If the link moves with zero angular acceleration, the acceleration of the mass centre being 10 m/s^2, find the magnitude of this force and the necessary additional torque which must act on the link.

12.3. Write a program to solve the kinematic and force analysis problem for a reciprocating pump as described in this chapter. Test your program by a graphical analysis for particular values, using the virtual-work method.

12.4. A scotch yoke mechanism is driven at 200 revs/min, the stroke is 300 mm and the slider has a mass of 25 kg. The motion is horizontal and friction can be ignored. Calculate the maximum instantaneous torque required at the driving link, on the assumption of a constant-speed motor.

13 GEAR-TRAINS AND DIFFERENTIALS

13.1   Transmission ratios

If two parallel shafts a and b rotate with speeds ω_a and ω_b respectively, the *transmission ratio R_{ab}* is defined as

$$R_{ab} = \omega_a \, / \, \omega_b$$

which is positive if the shafts rotate in the same direction and negative if they rotate in opposite directions. Thus the transmission ratio between two spur gears, or between two helical gears on parallel shafts, is given by

$$R_{ab} = -N_b/N_a$$

where the Ns are the numbers of teeth in each gear. As is well known, the transmission ratios of the individual gear-pairs are multiplied together to give the overall ratio of a train of gears. The train in Fig. 13.1, for example, has an overall transmission ratio of

$$R_{af} = (-N_f/N_e)(-N_d/N_c)(-N_b/N_a)$$

$$= R_{ef}R_{cd}R_{ab}$$

When the shafts are not parallel, say when bevel gears are used, the *magnitude* of the transmission ratio is obtained by the same method but the sign is not so obvious. It can be found easily by the following method. Draw on each shaft an arrow pointing in a direction such that you would be looking *along* the arrow if the rotation, as you viewed it, was clockwise. Fig. 13.2 shows this done for a pair of spur gears. Viewed in elevation the diagram appears as in Fig. 13.3(a). Now suppose the gears to be bevels with only a small angle, as in Fig. 13.3(b); the senses of rotations, and the directions of the arrows, are easily understood. Ex-

Fig. 13.1. Finding the transmission ratio of a gear-train.

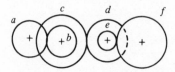

tending the idea, the diagram for a pair of 90° bevels appears as in Fig. 13.3(c). Thus the behaviour of a complex train can easily be understood by sketching the arrows, as in Fig. 13.4.

13.2 Synthesis of gear-trains

We wish to find the least number of gears that will provide, either exactly or within specified limits, a required overall transmission ratio. It is also good practice to ensure that the tooth-numbers in each meshing pair are prime to each

Fig. 13.2. Use of arrows to represent the sense of rotation.

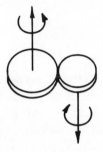

Fig. 13.3. Sense of rotation of bevel gears.

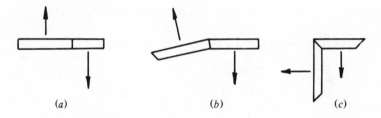

      (a)   (b)   (c)

Fig. 13.4. Use of arrows to find the sign of the transmission ratio of a complex gear-train.

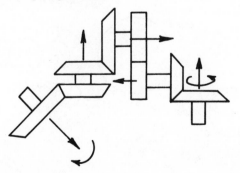

other, so as to distribute wear evenly, especially if the load fluctuates. Finally, there are practical limits on gear size to be taken into account.

Example 1. Find a train with a ratio of 20/27, using only external gears. No gear should have less than 25 or more than 180 teeth.

Since the ratio is positive, and external gears are to be used, we will need two meshings, i.e. four gears. Factorising numerator and denominator to obtain four numbers we have $20/27 = (4 \times 5)/(3 \times 9)$. Now we multiply each number, starting with the smallest, by the least factor which will make it equal to or greater than 25. Thus we must multiply 3 in the denominator by 9, and 4 in the numerator by 7. The remaining numbers must then be multiplied by factors such that all the factors cancel out, leaving the ratio unchanged. This gives

$$R = \frac{(7)4 \times (9)5}{(9)3 \times (7)9} = \frac{28 \times 45}{27 \times 63}$$

giving the train shown in Fig. 13.5(*a*). If it had turned out that all the pairs of tooth-numbers were prime to each other that would complete the design. Unfortunately, in this example, 45 and 63 have the common factor 9; and the equivalent arrangement, $(28 \times 45)/(63 \times 27)$, has the same fault, since 45 and 27 also have 9 as a common factor. We can overcome this difficulty by using a pair of idlers, as in Fig. 13.5(*b*). Two are needed, to keep *R* positive, and they have been chosen so that they are prime to each other and to the gears they mesh with. But we could (and should) achieve the same result more simply as in Fig. 13.5(*c*). This

Fig. 13.5. Synthesis of a gear-train for specified transmission ratio.

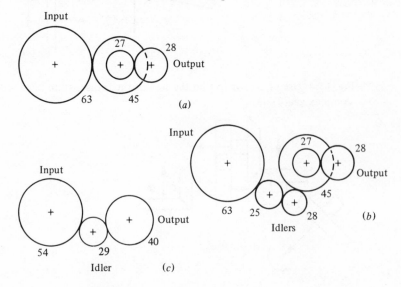

example shows that, in general, there are several solutions to a gear-train synthesis problem.

Example 2. In an astronomical clock we need a gear-train giving the ratio of a year to a day, correct to one second if possible. This ratio is 164 359/450 which can be factorised to (269 x 26 x 94)/(10 x 10 x 18). Unfortunately 269 is prime, and too large for a practical train. But 269 001/1000 = 269 to one part in 269 000 and can be factorised to give (81 x 81 x 41)/(100 x 10). Thus an approximate solution (to within 2 minutes per year) is given by $R = (26 \times 94 \times 81 \times 81 \times 41)/(10 \times 10 \times 18 \times 100 \times 10)$. In clockwork, and other very lightly loaded slow-running machines, small tooth-numbers and common factors between meshing gears are acceptable. But the solution is crude and cumbersome.

For problems of this sort it is essential to use more advanced methods, outside our scope. An account of them is given by H. E. Merritt, *Gear Trains*, Pitman, 1947, who also provides tables of prime factors. Using these methods we can approximate to the required ratio by (48 x 89 x 97)/(8 x 10 x 13) which uses six gears instead of ten and is accurate to 7.4 seconds per year.

13.3 Reverted trains

A *reverted train* is defined as one in which the input and output shafts are co-axial; a simple example is shown in Fig. 13.6(*a*). The transmission ratio of such a train is, of course, found in the same way as for any other train and it can be syn-thesised by the same methods. One advantage of a reverted train is that we can use several intermediate gears, as in Fig. 13.6(*b*), thus increasing the amount of power that can be transmitted by a train of given overall size. Its main use, how-ever, is in epicyclic gearing (also called planetary gearing) and with this in view we introduce the following definitions, that will be necessary for the next section.

The two co-axial shafts of a reverted train are called *central shafts* and the gears on them *central gears*. The other gears are called *planets*. The frame carrying the planets is called the *arm* (some writers call it the carrier, or even the spider). Suppose the arm to be fixed and the train to be used as an ordinary gear-train.

Fig. 13.6. Reverted train, and use of multiple planets.

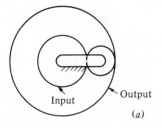

(a)

(b)

The transmission ratio *with the faster of the central shafts as input* is then called the *basic ratio* and we shall denote it by ρ. Thus defined, the basic ratio must be greater than $+1$, or equal to or less than -1. It is physically possible for it to be exactly $+1$, but then the train would act simply as a rigid coupling between the central shafts, so this case is of no importance.

Some examples of reverted trains are shown (with single planets for clarity) in Fig. 13.7. Those shown in (a), (b), (c) and (h) are the commonest types.

13.4 Epicyclic and differential trains

An epicyclic train is a reverted train in which the arm rotates. One of the central shafts may be fixed, in which case power is transmitted between the rotating arm and the other central shaft. We can, however, use the mechanism in such a way that the arm and both central shafts rotate. When this happens the train is called a differential. The differential used on the back axle of a motor vehicle is an example. Two or more epicyclic trains can be connected together to form a compound train.

Fig. 13.7. Reverted trains.

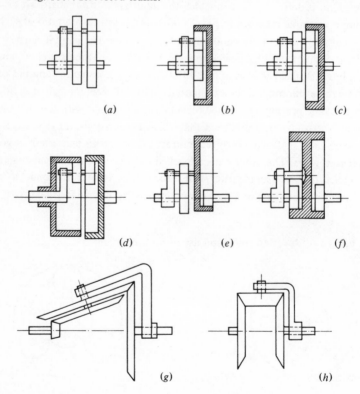

(a) (b) (c)

(d) (e) (f)

(g) (h)

The detailed kinematic behaviour of an individual epicyclic train is best studied by the instantaneous-centre method (Chapter 4). There is, however, a general formula that applies to any epicyclic train without regard to its internal construction, and is useful for the analysis and synthesis of compound trains. We shall see that an epicyclic can be regarded as a closed box with three co-axial shafts, two of them the central shafts and the third being connected to the arm; and that the relation between the speeds of those three shafts depends solely on the basic ratio ρ.

Consider such a train as shown in Fig. 13.8. We mark the angular positions of the central shafts by lines 1 and 2 (1 being the faster shaft if the arm were fixed); the angular position of the arm is marked by line 3. We measure angles *relative to the arm* by ϕ_1 and ϕ_2, and angles *relative to the fixed datum* by θ_1, θ_2 and θ_3. Now it is obvious that $\phi_1 = \theta_1 - \theta_3$ and $\phi_2 = \theta_2 - \theta_3$. But the motion of the central shafts *relative to the arm* is not affected by the fact that the arm itself is rotating. Hence $\phi_1/\phi_2 = \rho$, as defined in Section 13.3. Thus we have $(\theta_1 - \theta_3)/(\theta_2 - \theta_3) = \rho$ and hence $\theta_1 - \theta_2 + (\rho - 1)\theta_3 = 0$. Finally, differentiating with respect to time to get angular velocities, we have $\dot{\theta}_1 - \rho\dot{\theta}_2 + (\rho - 1)\dot{\theta}_3 = 0$.

For convenience in later working we replace the three angular velocities $\dot{\theta}_1$, $\dot{\theta}_2$ and $\dot{\theta}_3$ by the symbols p, q and r respectively. The formula therefore becomes

$$p - \rho q + (\rho - 1) r = 0 \qquad\qquad (13.1)$$

Since a differential combines two rotations to give a single output it has many applications. For example, the mechanism shown in Fig. 13.9 has $\rho = -1$ and therefore $p + q - 2r = 0$. Hence $r = (p + q)/2$. Now if we drive the central shafts at equal speeds but in opposite directions we will have $r = 0$, meaning that the arm will be stationary; and if the speeds vary r will be proportional to the difference between them, so that the arm will start to rotate. Hence the mechanism

Fig. 13.8. Shaft rotations of a differential, measured relative to the arm and relative to a fixed datum.

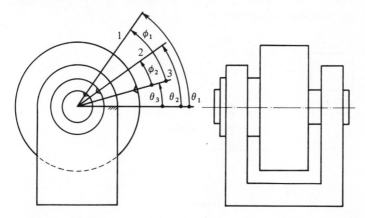

can be used in a control system to keep the two central shafts at the same speed. It has been used in this way to match the speed of a water turbine against that of a small constant-speed reference motor, the rotation of the arm being used to control the water supply to the turbine.

Another use of a differential is to combine the speeds of two electric motors (Fig. 13.10). Direct speed variation of a large motor is expensive so a differential is used to combine the speed of a large constant-speed motor with that of a small, easily-controlled variable-speed motor. In the example $\rho = -5$; substituting this value into equation (13.1) we see that $r = 5q/6 + p/6$, where r is the output speed, q the speed of the large motor and p the speed of the small motor.

Instead of using the epicyclic train as a differential we can use it as an ordinary transmission with a single input and a single output, provided we lock one

Fig. 13.9. Differential used as an element in a control system. The output shaft is stationary so long as the input shafts have equal and opposite speeds.

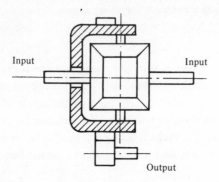

Fig. 13.10. Differential used to combine the speeds of two motors, giving a variable speed drive.

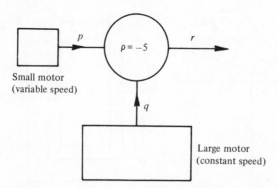

of the co-axial shafts. If we lock the arm we get an ordinary (but reverted) gear-train. If we lock central shaft 1 we must put $p = 0$ in equation (13.1). Taking the arm as output and the other central shaft as input we have

$$R_{23} = \frac{q}{r} = \frac{\rho - 1}{\rho} \tag{13.2}$$

Taking the arm as input and the other central shaft as output we have

$$R_{32} = \frac{r}{q} = \frac{\rho}{\rho - 1} \tag{13.3}$$

Obviously R_{32} will be very large if ρ is only slightly greater than one. Suppose, for example, we want a transmission with $R = 100$. Putting this value of R into equation (13.3) and solving for ρ, we get $\rho = 100/99$. Since this is positive we could use type (a) in Fig. 13.7, giving the mechanism shown in Fig. 13.11(a). Unfortunately, for reasons that will appear in the next chapter, this design would have a very low efficiency; a better solution would be to use type (d), as shown in Fig. 13.11(b).

The other inversion is that in which central shaft 2 is locked. If we use the arm as output this gives

$$R_{13} = 1 - \rho \tag{13.4}$$

and if the arm is used as input

$$R_{31} = \frac{1}{1 - \rho} \tag{13.5}$$

the last case giving large negative values of R if ρ is slightly greater than one.

It is important to remember that for large values of R, either positive or negative, obtained with small values of ρ, a very small *change* in ρ will give a large change in R. This can easily be seen by differentiating equations (13.3) and (13.5) with respect to ρ. This means that a high-ratio epicyclic using friction gears will be unreliable. If toothed gearing is used small fluctuations in basic ratio

Fig. 13.11. High-ratio single-stage epicyclic transmissions. Type (a) has a very low efficiency, type (b) is better.

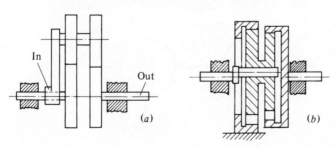

caused by profile errors in the teeth will give large fluctuations in transmission
ratio, and hence impose high inertial loads. Good quality gears and careful atten-
tion to detail design are essential if a large transmission ratio must be achieved in
a single stage. It is obviously better to use a compound train if the extra size,
weight and cost can be tolerated.

13.5   Compound trains

The simplest type of compound is that in which there is only one connection between each pair of epicyclics, as in Fig. 13.12(*a*). We can represent the train sym-
bolically by the simple diagram shown in Fig. 13.12(*b*), in which each epicyclic
is represented by a circle with its basic ratio written inside. The transmission ratio
of the whole train is the product of the transmission ratios of the individual epicyclics. In this example we have

$$R = R_1 R_2$$

$$= (1 - \rho_1)\left(\frac{\rho_2 - 1}{\rho_2}\right)$$

$$= (1 + 9/5)\left(\frac{-19/9 - 1}{-19/9}\right)$$

$$= 392/95$$

Fig. 13.12. Finding the transmission ratio of a compound train with a
single connection between the two epicyclics.

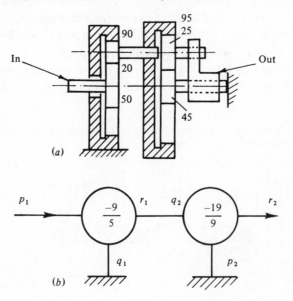

(a)

(b)

Sometimes we meet an arrangement as in Fig. 13.13, in which one of the central gears and a planet are missing; this is called a *united train*. In this example gears 1, 2 and 3 act in the usual way as the left-hand differential, while gears 1, 2, 4 and 5 act as the right-hand differential. The mechanism is equivalent to the ordinary compound shown in Fig. 13.13(*b*).

To see how we can analyse compounds with more than one connection between a pair of epicyclics, let us consider the example shown in Fig. 13.14. We

Fig. 13.13. A united train and the equivalent compound.

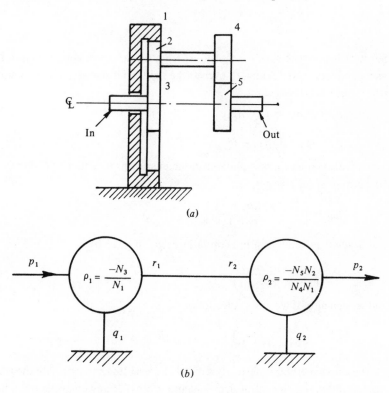

(*a*)

(*b*)

Fig. 13.14. Finding the transmission ratio of a compound train with multiple connections.

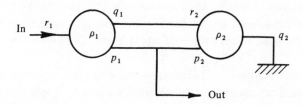

write equation (13.1) for each separate epicyclic, and further equations expressing the fact that certain shafts are locked (having zero speed) and that certain
pairs of shafts have equal speeds because they are connected together. In this
example the equations are

$$p_1 - \rho_1 q_1 + (\rho_1 - 1) r_1 = 0$$

$$p_2 - \rho_2 q_2 + (\rho_2 - 1) r_2 = 0$$

$$q_2 = 0$$

$$r_2 = q_1$$

$$p_2 = p_1$$

Substituting the values of q_2, r_2 and p_2 from the last three equations into the
second equation, and taking r_1 over to the right-hand side as a known quantity
(the input speed) we get

$$p_1 - \rho_1 q_1 = (1 - \rho_1) r_1$$

$$p_1 + (\rho_2 - 1) q_1 = 0$$

These are linear algebraic equations in the unknowns p_1 and q_1. Solving for p_1
(the output speed) we get

$$p_1 = \left[\frac{(1 - \rho_1)(\rho_2 - 1)}{\rho_1 + \rho_2 - 1} \right] r_1$$

and hence the overall transmission ratio, r_1/p_1, is given by

$$R = \frac{\rho_1 + \rho_2 - 1}{(1 - \rho_1)(\rho_2 - 1)} \tag{13.6}$$

and we can also, if we wish, solve for q_1 to give

$$q_1 = \left(\frac{\rho_1 - 1}{\rho_1 + \rho_2 - 1} \right) r_1$$

The above method is perfectly general and enables us to calculate the transmission ratio (and all central shaft and arm speeds) of any epicyclic train, however complex. To summarise: sketch the train in symbolic form and label the
shafts p_1, q_1 and r_1 for epicyclic 1, p_2 etc. for epicyclic 2 and so on. Write out
equation (13.1) for each epicyclic. Make the substitutions indicated by the
lockings and connections. Solve the equations. If the system contains a united
train this must be replaced by the equivalent separate epicyclics before starting
the analysis.

Methods based on equation (13.1) cannot, of course, give any information
about planet speeds. That equation, as we saw when we derived it, takes no ac-

count of the internal construction of the individual epicyclics. But once we have found the central shaft and arm speeds by our present method we can, if necessary, find the planet speeds by the instantaneous-centre method (or the tabular method). This requires a knowledge of the internal construction of each epicyclic.

13.6 Synthesis of epicyclic trains

The synthesis of a single epicyclic is a simple matter. We wish to connect up three shafts so that they have specified speeds ω_1, ω_2 and ω_3. Now from equation (13.1), solving for ρ, we have

$$\rho = (r - p)/(r - q) \tag{13.7}$$

The epicyclic can be connected to the three shafts in six different ways, giving six different values of ρ, according to the following table

p	q	r	ρ
ω_1	ω_2	ω_3	$(\omega_3 - \omega_1)/(\omega_3 - \omega_2)$
ω_1	ω_3	ω_2	$(\omega_2 - \omega_1)/(\omega_2 - \omega_3)$
ω_2	ω_3	ω_1	$(\omega_1 - \omega_2)/(\omega_1 - \omega_3)$
ω_2	ω_1	ω_3	$(\omega_3 - \omega_2)/(\omega_3 - \omega_1)$
ω_3	ω_1	ω_2	$(\omega_2 - \omega_3)/(\omega_2 - \omega_1)$
ω_3	ω_2	ω_1	$(\omega_1 - \omega_3)/(\omega_1 - \omega_2)$

It can easily be seen (by inspection of the table) that the last three values of ρ are reciprocals of the first three. Since ρ must be numerically greater than or equal to unity, it follows that there will only be three distinct solutions. We have assumed here that the epicyclic is used as a differential, but the result can of course be applied to an ordinary transmission, where one of the specified speeds is zero.

To synthesise a transmission using a compound train we must first carry out the sort of analysis illustrated by the example (Fig. 13.14) in Section 13.5, and then substitute values of the basic ratios into the formula for the transmission ratio. We see from equation (13.6), for example, that either of the two basic ratios can be chosen freely, the other then being given by the equation for a specified value of R.

13.7 Epicyclic gearboxes

A compound epicyclic train can be used as a gearbox, gear-changing being done by applying brakes. In this way we can avoid the use of sliding gears or clutches. Consider the two-speed box shown in Fig. 13.15. When the right-hand brake is applied we have the mechanism of Fig. 13.14, whose transmission ratio is given by equation (13.6) above. When the left-hand brake is applied the power is taken entirely through train 1, and the other train idles. In this second case the transmission ratio is given by

$$R' = 1/(1 - \rho_1) \tag{13.8}$$

Now suppose we want to design such a box to give specified ratios R with the right-hand brake on and R' with the left-hand brake on. We have to find the basic ratios ρ_1 and ρ_2. From equation (13.8) we have

$$\rho_1 = 1 - 1/R' \tag{13.9}$$

and we now substitute this value of ρ_1 into equation (13.6), solving for ρ_2 to get

$$\rho_2 = (R - 1)/(R - R') \tag{13.10}$$

In general, to design an n-speed gearbox we need n differentials and these must be arranged in a train which, with all the brakes off, has mobility $M = 2$. Applying a brake provides one of the inputs, namely zero speed, and hence only one other input is needed to drive the train and transmit power.

13.8 Systematics of compound trains

It may be asked if the layout in Fig. 13.15 has to be arrived at by an inspired guess, or can be reached by some logical process. Is it the only possible layout for a two-speed box; and if not, what are the other possibilities? Moreover, suppose we want to design a box with more than two speeds. We cannot apply the usual methods of systematics, since we are no longer dealing with kinematic pairs; our basic elements are the individual differentials and each of them has three shafts.

Let us consider what happens when we assemble a compound. We start with n separate differentials, so we have $3n$ variables (three shaft speeds for each) and n

Fig. 13.15. A two-speed epicyclic gearbox. When the left-hand brake is engaged the left-hand epicyclic acts as a single-stage transmission, the right-hand epicyclic idling. When the right-hand brake is engaged we have the compound train of Fig. 13.14.

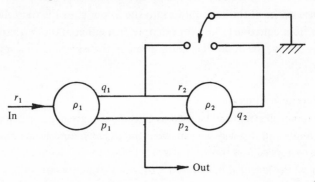

equations of type (13.1), one for each differential. Now each connection between two shafts introduces another equation. For example, if we connect central shaft 1 on train 2 to the arm on train 3 we will have the additional equation $p_2 = r_3$. It follows that if there are c connections we will have another c equations. Also, each locked shaft introduces another equation. For example, if we lock central shaft 2 on train 3 we will have $q_3 = 0$. So if there are l locked shafts we will have a further l equations. Hence there are, altogether, $3n$ variables related by $(n + c + l)$ equations.

Now some of these variables (shaft speeds) will be known quantities, the input speeds, and the others will be the unknowns. But if the equations are to have solutions there must be just as many unknowns as equations. Hence the number of inputs, i.e. the mobility M, must equal the difference between the number of variables and the number of equations, so $M = 3n - (n + c + l)$. This reduces to

$$M = 2n - c - l \tag{13.11}$$

The analysis carried out above for the mechanism shown in Fig. 13.14 shows how the theory works in a specific case. There we had two differentials ($n = 2$) and hence $3n = 6$ variables. We wrote two equations of type (13.1), one for each differential, two more since we had two connections, and lastly a fifth equation to express the fact that one shaft was locked. Since we had six variables but only five equations it followed that one variable had to be a known quantity, i.e. that the train had a single input. Thus we ensured that the number of equations was the same as the number of unknowns and hence that the set of equations could be solved, as we verified by actually solving them. We can also verify equation (13.11) by applying it directly to this example without actually writing, let alone solving, any equations. There are two differentials, two connections and one locking, so $M = 2 \times 2 - 2 - 1 = 1$.

For use in type synthesis we re-arrange equation (13.11) as

$$c + l = 2n - M \tag{13.12}$$

so that we can enumerate the class of compound trains with n differentials and mobility M. Let us see how this is done by using the method to find all possible two-speed epicyclic boxes. We have $n = 2$ and $M = 2$, so $c + l = 2 \times 2 - 2 = 2$. So we have to consider all possible combinations of connections and lockings that add up to 2.

This is easily done. There must be at least one connection if the two differentials are to form a system at all. Obviously there cannot be less than zero lockings. So there are only two possible arrangements, which we set out in the following table

c	l
1	1
2	0

These two possible compounds are shown in Fig. 13.16. Finally, we must consider the various possible ways of applying the brakes and making the input and output connections. This is done by trial; we soon discover that only the second compound can be used, and that it provides the three possible arrangements shown in Fig. 13.17. A choice of the most suitable type depends on analysis of the efficiencies of the various possible *internal* arrangements for each differential, using the methods described in the next chapter.

The same method can be used to enumerate more complex trains, but slightly more care is needed because of the numerous ways in which the connections and lockings can be arranged. Consider the problem of designing a three-stage single speed transmission, where we need $M = 1$ and $n = 3$. Equation (13.12) gives $c + l = 5$ and hence the following table:

c	l
2	3
3	2
4	1

Fig. 13.16. Two-stage compounds with $M = 2$.

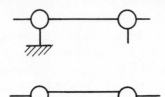

Fig. 13.17. The possible basic arrangements of a two-speed gearbox.

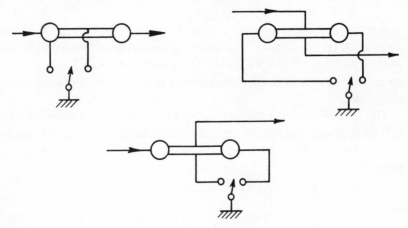

Each row in the table gives several possible arrangements, as shown in Fig. 13.18. Note that the mechanism shown in broken lines has to be excluded as the two left-hand differentials form a compound with $M = 0$, reducing the system to a single-stage transmission with only the right-hand differential working; the compound would certainly function, but it is of no practical value.

13.9   Other differential mechanisms

Equation (13.1) was deduced from purely geometrical properties of Fig. 13.8. It does not depend on there being gear-wheels or any other particular mechanism in the 'closed box' with its three co-axial shafts. The only assumption we make is that the three shafts are co-axial and that, inside the box, there is some sort of connection between them. It follows that we can use devices other than gears to make a differential. We could, for example, connect the shafts together by belts or chains, or we could use a ball-race as a friction drive differential for light loads, the cage acting as the arm and the balls as planets.

Finally, the whole theory can be applied as it stands to mechanisms that have linear parallel motion instead of co-axial rotary motion. The geometry of such a device is shown in Fig. 13.19. We use the same notation as in Fig. 13.8, but now the symbols stand for lengths instead of angles; and by the same reasoning as we used before, we arrive again at equation (13.1), where this time the symbols stand for linear instead of angular velocities. All the rest of the theory then follows exactly as before, since everything was founded on that equation alone.

The theory can then be applied to lever systems, differential screws and systems of cords and pulleys. For example, Fig. 13.20 shows the Chinese windlass

Fig. 13.18. Three-stage epicyclic transmissions.

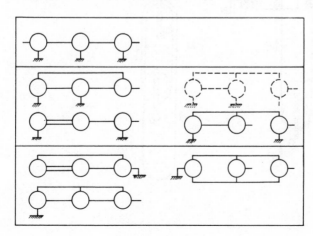

and its modern form, Weston's pulley block. The hanging pulley acts as a differential of basic ratio $\rho = -1$, where we consider the hook as the arm, and combines the two slightly different linear velocities of the ropes to give a very slow arm velocity and hence a high mechanical advantage.

Fig. 13.19. A differential giving linear instead of rotary motion. This figure should be compared with Fig. 13.8, showing the rotary case.

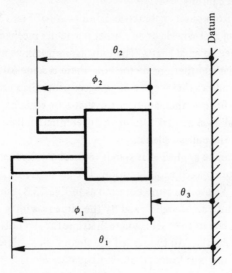

Fig. 13.20. The Chinese windlass and Weston's hoist, both using a moving pulley as a differential.

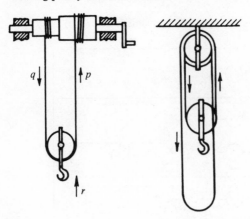

Exercises 13

13.1. Find the transmission ratio of the gear-train shown in Fig. 13.21.

Fig. 13.21.

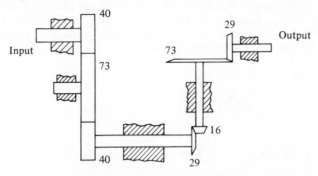

13.2. Select tooth-numbers for a train of spur gears with a transmission ratio of 730/261. No gear is to have less than 25 teeth or more than 100 and no meshing pair is to have a common factor.

13.3. Sketch a train having the least possible number of gears with a transmission ratio of 235/185, no gear having less than 25 teeth or more than 125.

13.4. Check your solution to Exercise 4.7 by the methods of this chapter (equation (13.1)).

13.5. Calculate the basic ratio of the train shown in Fig. 13.22 and hence find its transmission ratio when it is used with the left-hand shaft locked, the arm as input and the right-hand shaft as output.

Fig. 13.22.

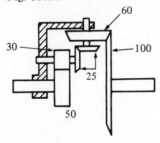

13.6. The three shafts of a differential are to have speeds of 400, 300 and − 180 revs/min, but it is not specified which shaft is to have which speed. Show that there are three (and only three) possible values for the basic ratio. Choose one, giving reasons for your preference, and sketch the train showing tooth-numbers.

13.7. Check your solution to Exercise 4.8 by the methods of this chapter.

13.8. Fig. 13.23 shows a train in which the input shaft is keyed to gear A and the output shaft to gear D, these gears being free to rotate relative to the arm. Gears B and C are keyed together but are also free to rotate relative to the arm. Gear E is the fixed member. Identify the distinct simple differentials, find their basic ratios and sketch a symbolic diagram of the train. Calculate its transmission ratio.

Fig. 13.23.

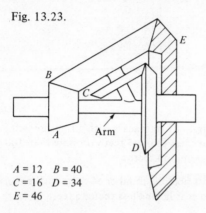

$A = 12$ $B = 40$
$C = 16$ $D = 34$
$E = 46$

13.9. Fig. 13.24 shows a two-speed epicyclic transmission, required to have transmission ratios of 4 : 1 with q_1 connected to r_2 and 3 : 1 with q_1 locked and r_2 left free to idle. Calculate the basic ratios ρ_1 and ρ_2. What additional feature is needed so that the transmission can also provide direct drive, i.e. a ratio of 1 : 1?

Fig. 13.24.

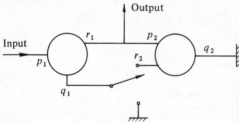

13.10. Fig. 13.25 shows a machine vice worked by a differential screw. The small thread has a pitch of 2 mm, the large thread a pitch of 3 mm. Sketch an analogous epicyclic train. (Hint: take the screw as fixed member when finding the basic ratio.) Hence calculate the distance by which the vice closes when the handle is turned through one revolution.

Fig. 13.25.

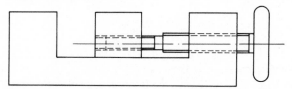

13.11. Fig. 13.26 shows a pulley train used as a hoist. Sketch an analogous epicyclic train, and find the distance through which the rope must be pulled to raise the load by one metre.

Fig. 13.26.

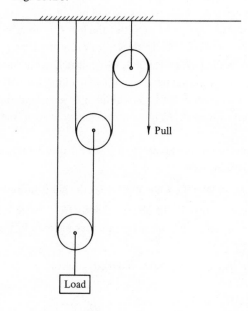

14 EFFICIENCY OF GEAR-TRAINS AND DIFFERENTIAL MECHANISMS

14.1   Gear-train efficiency

Ordinary gear-trains are highly efficient, the losses at each meshing being only about 1% of the transmitted power. The power loss in an ordinary train is therefore not important unless large power is to be transmitted, in which case we can sometimes face a difficult problem in getting rid of the heat.

The efficiency E of a pair of meshing gears is given by $E = 1 - L$, where L is the loss. Both E and L are *fractions* of the *input* power, not percentages. If we have a train with two meshings the overall efficiency will be $E = E_1 E_2$. Now $E_1 E_2 = (1 - L_1)(1 - L_2) = 1 - (L_1 + L_2) + L_1 L_2$. But the loss L is usually quite small, say 0.01 or 0.02, so $L_1 L_2$ will be negligible. If, for example, $L_1 = L_2 = 0.02$, then $(L_1 + L_2) = 0.04$, while $L_1 L_2 = 0.0004$, which is negligibly small compared with $(L_1 + L_2)$. Thus, within practical limits of accuracy, $E = 1 - (L_1 + L_2)$. The argument is easily extended to any number of gears and therefore, in general, $E = 1 - $ (sum of losses). We must now see how we can find the value of L.

The theory of gear efficiency is quite complex, but for most ordinary design purposes we can use the following rules:

(a) For *external* spur gears take the value of L from the chart in Fig. 14.1.

(b) For *internal* spur gears multiply the above value by $(R - 1)/(R + 1)$ where $R = $ (number of teeth in large gear)/(number of teeth in small gear).

We see that the loss can be greatly reduced, when R is small, by using internal gears. For example, using gears of 92 and 100 teeth the chart gives $L = 0.0048$; but if we use internal gears,

$$L = 0.0048 \left(\frac{100}{92} - 1 \right) \Big/ \left(\frac{100}{92} + 1 \right) = 0.0002,$$

reducing the loss to a mere 1/24 of the external-gear value. If we were transmitting 1000 h.p. the external gears would give a loss of 3.1 kW, a considerable amount of heat to get rid of, but the use of internal gears would reduce the loss to 0.13 kW.

(c) For helical gears the value of L given by the chart must be multiplied by $0.8 \cos \sigma$ where σ is the spiral angle. If they are internal gears we must also multiply by $(R - 1)/(R + 1)$.

(d) For bevel gears the number of teeth in each wheel must be multiplied by $\sec \theta$, where θ is the pitch-cone angle, to give the 'virtual tooth-numbers'. These virtual tooth-numbers are then used, instead of the actual tooth-numbers, to find L from the chart.

Within the limits of accuracy considered when forming these working rules, L is the same whichever wheel is the driver and hence E does not depend on the direction of power flow through the train; and our calculations automatically include a 'reasonable' allowance for bearing friction, of course on the assumption of 'reasonable' bearing design and quality. It cannot be emphasised too strongly that calculated values of loss and efficiency, on the basis of these rules, are only approximate. The actual values depend on speeds, loads, gear material and

Fig. 14.1. Chart for estimating the power loss, as a fraction of the input power, in a pair of external spur gears. (Based on data by H. E. Merritt, *Gear Trains*, Pitman 1947).

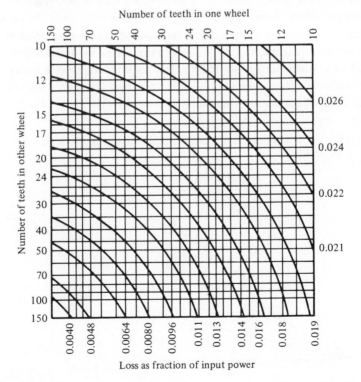

Number of teeth in one wheel

Loss as fraction of input power

finish, shaft alignment and rigidity, and properties of the lubricant, in ways that cannot be predicted theoretically with any great accuracy. We should use the calculations to compare alternative proposed designs, rather than to develop an existing unit. For the latter purpose actual tests of efficiency under working conditions are essential.

14.2 Epicyclic gear efficiency

Epicyclic trains behave in peculiar ways. A properly designed epicyclic can be more efficient than an ordinary train using the same gears; if the design is wrong the losses can be so great that the train will not run at all.

Referring to Fig. 14.2, we denote the torques applied to central shaft 1, central shaft 2 and the arm by P, Q and R respectively. We shall adopt the same approach that we used for the kinematics in Chapter 13: we consider what happens *relative to the arm*, since the internal power flow and losses cannot be affected by rotation of the whole unit. All that matters is rotation of the central shafts relative to the arm, as measured by the angles ϕ_1 and ϕ_2; for this is the only motion that can cause relative movement of contacting teeth, and hence transmit power or waste it in friction. Just as we worked out the kinematics in terms of the basic ratio ρ, which is the transmission ratio of the train used as an ordinary gear-train, so we shall work out efficiencies in terms of the *basic efficiency E_0*, which is the efficiency of the train when used as an ordinary gear-train with the arm fixed. We can, of course, calculate E_0 by the rules given in Section 14.1. When doing so we count only a single planet, however many there are in practice, on the assumption that the power flow (and hence the loss) is equally divided between the planets.

Fig. 14.2. Torques, and rotation angles relative to the arm, in a differential.

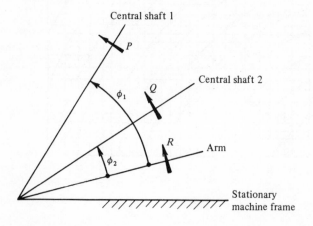

We take all the torques and angular velocities as positive if they are counter-clockwise, as shown in the figure. Hence, for any member of the train, the product of torque and angular velocity will be a power input if it is positive (torque and angular velocity in the same sense) but a power output if it is negative.

We assume that the train is running at constant speed, or at least that the speed changes so slowly that changes in the internal kinetic energy of the train can be ignored. Then for static equilibrium

$$P + Q + R = 0 \tag{14.1}$$

and for power balance

$$E_0 \dot{\phi}_1 P + \dot{\phi}_2 Q = 0 \tag{14.2}$$

if the input is to central shaft 1 and the output from central shaft 2; or

$$\dot{\phi}_1 P + E_0 \dot{\phi}_2 Q = 0 \tag{14.3}$$

if the input is to central shaft 2 and the output from central shaft 1.

Now if the direction of power flow is from central shaft 1 to central shaft 2 we must solve simultaneously equations (14.1) and (14.2); if the other way, equations (14.1) and (14.3). Remembering that $\dot{\phi}_1/\dot{\phi}_2 = \rho$, the solutions are

Power flow $1 \rightarrow 2$

$$P = R/(\rho E_0 - 1) \tag{14.4}$$

$$Q = -\rho R E_0/(\rho E_0 - 1) \tag{14.5}$$

Power flow $2 \rightarrow 1$

$$P = E_0 R/(\rho - E_0) \tag{14.6}$$

$$Q = -\rho R/(\rho - E_0) \tag{14.7}$$

The application of these equations is as follows. We have made no assumption about any shaft (including the arm) being fixed or not, so our results apply to any use of the train, provided we can find out which way the power flows *relative to the arm*. Consider first the case where one central shaft is fixed. The other must be specified as either input or output shaft, which means we know the sign of the torque applied to it. Now from the kinematics we know the sign of the angular velocity of that shaft relative to the arm, so we can see which way the power flow occurs.

There are in fact eight possible cases; all of them are tabulated in Fig. 14.3. These tabulated formulae can safely be used without an understanding of the theory. But, for the reader who wishes to dip a little deeper into this rather difficult topic, the following explanation of the working of the case with shaft 1 fixed and shaft 2 as input may be of interest.

Since $p = 0$ we have $q/r = (\rho - 1)/\rho$. Now $\dot{\phi}_2 = q - r$, but $r = \rho q/(\rho - 1)$, so *if ρ is positive* we have $r > q$. Thus $q - r$, and hence $\dot{\phi}_2$, are negative. Hence the power flow relative to the arm is from shaft 1 to shaft 2, and we must use equations (14.4) and (14.5). The latter gives us the input torque, $Q = - \rho R E_0 / (\rho E_0 - 1)$. Now the power flow *into* the train at the output shaft is $r R$, remembering our sign convention for speeds and torques. But to find the efficiency we need the *output* power, and this will be $- r R$. The input power is $q Q = - q \rho R E_0 / (\rho E_0 - 1)$. Hence efficiency = (output power)/(input power) = $(\rho E_0 - 1)/E_0 (\rho - 1)$. Note that we would have a different result if the basic ratio were

Fig. 14.3. Efficiency and torques in an epicyclic train.

	Fixed shaft	Input shaft	Transmission ratio	P	Q	R	Efficiency
Positive basic ratios	2	1	$1 - \rho$	$\dfrac{-R}{1 - \rho E_0}$	$\dfrac{\rho E_0 R}{1 - \rho E_0}$	R	$\dfrac{\rho E_0 - 1}{\rho - 1}$
	2	Arm	$\dfrac{1}{1 - \rho}$	P	$\dfrac{-\rho P}{E_0}$	$\left(\dfrac{\rho - E_0}{E_0}\right) P$	$\dfrac{E_0 (\rho - 1)}{\rho - E_0}$
	1	2	$\dfrac{\rho - 1}{\rho}$	$\dfrac{R}{\rho E_0 - 1}$	$\dfrac{-\rho E_0 R}{\rho E_0 - 1}$	R	$\dfrac{\rho E_0 - 1}{E_0 (\rho - 1)}$
	1	Arm	$\dfrac{\rho}{\rho - 1}$	$-\dfrac{E_0}{\rho} Q$	Q	$-\left(\dfrac{\rho - E_0}{\rho}\right) Q$	$\dfrac{\rho - 1}{\rho - E_0}$
Negative basic ratios	2	1	$1 - \rho$	$\dfrac{-R}{1 - \rho E_0}$	$\dfrac{\rho E_0 R}{1 - \rho E_0}$	R	$\dfrac{\rho E_0 - 1}{\rho - 1}$
	2	Arm	$\dfrac{1}{1 - \rho}$	P	$\dfrac{-\rho P}{E_0}$	$\left(\dfrac{\rho - E_0}{E_0}\right) P$	$\dfrac{E_0 (\rho - 1)}{\rho - E_0}$
	1	2	$\dfrac{\rho - 1}{\rho}$	$\dfrac{R E_0}{\rho - E_0}$	$\dfrac{-\rho R}{\rho - E_0}$	R	$\dfrac{\rho - E_0}{\rho - 1}$
	1	Arm	$\dfrac{\rho}{\rho - 1}$	$\dfrac{-Q}{\rho E_0}$	Q	$-\left(\dfrac{\rho E_0 - 1}{\rho E_0}\right) Q$	$\dfrac{E_0 (\rho - 1)}{\rho E_0 - 1}$

Fig. 14.4. A high-ratio single-stage epicyclic transmission. This is a poor design, giving very low efficiency.

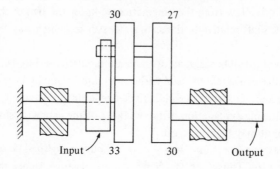

Input 33   30   Output

negative, because we would have a different direction of internal power flow. The example illustrates the distinction between external power flow, in this case from shaft 2 to the arm, and internal power flow, in this case from shaft 1 to shaft 2. Similar reasoning is used to establish the other results in the table.

In every tabulated case one shaft is fixed but we shall see, in the fourth of the following examples, how the tables can be used when both shafts, as well as the arm, are rotating.

Example 1. Find the efficiency of the high-ratio train shown in Fig. 14.4.

From Fig. 14.1 we get the basic efficiency, $E_0 = 1 - (0.014 + 0.013) = 0.97$. The basic ratio is $100/99$, which is positive, the arm is used as input, shaft 2 as output and shaft 1 is fixed. From the table the efficiency is $(100/99 - 1)/(100/99 - 97/100) = 0.25$, i.e. 25%. Clearly the mechanism is impracticable.

Example 2. As an alternative to the mechanism of Example 1 consider that shown in Fig. 14.5, which uses internal gears to obtain the same basic ratio and hence transmission ratio. We now have $E_0 = 0.9995$, and the efficiency is 0.95. Such a design would be adequate for low power, where the 5% loss would not cause a heat dissipation problem.

Example 3. An epicyclic train has basic ratio $- 2$ and basic efficiency 0.96. Central shaft 2 is locked and the arm is used as input member. Find the efficiency.

In this case the table tells us that the efficiency is 0.97, which is better than the basic efficiency.

Example 4. A load torque of 40 000 lbf in is to be driven at 1200 revs/min by the arm of a differential with $\rho = -3$ and $E_0 = 0.98$. A motor running at 1700

Fig. 14.5. An improvement on the design shown in Fig. 14.4.

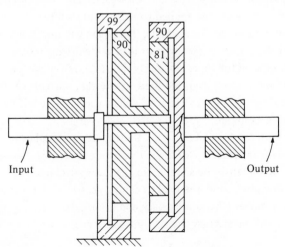

revs/min drives central shaft 2. Find the speed with which another motor must drive the remaining shaft, the motor torques, motor powers and the efficiency with which the train is working.

We are given $R = -40\,000$ (negative because this is the output shaft), $q = 1700$ and $r = 1200$. From the kinematics, $p = -300$ revs/min. Now the formulae in the table all require one of the shafts to be fixed, so we 'fix' a shaft by deducting its speed from that of all the shafts. Which shaft are we to fix? If we look at the table we can see that the formulae for torques are always given in terms of the load torque, since this is the only known torque in any of these problems. So in this case, since the load is applied to the arm, we can fix either of the central shafts. Let us fix shaft 1, by subtracting -300 revs/min from all the shaft speeds. Since this increases r to 1500, and R is negative, the arm remains the output member; whereas if we fixed shaft 2, it would not. Since ρ is negative we have established that the equivalent case is represented by the last but one line in the table. This gives $P = 9850$ lbf in. Since p is negative we have a power *output* at this shaft; the motor acts as a regenerative brake, absorbing 47 h.p. We also have $Q = 30\,200$ lbf in, so the input power is 813 h.p. The overall efficiency of the system depends on motor efficiencies and other electrical matters, particularly on the efficiency with which the small motor returns power to the line. The efficiency of the differential itself, however, is easily calculated. We should not use the horsepower figures, since they will have slight errors due to rounding off; it is better to use products of speed and torque directly. These tell us that efficiency = (output power)/(input power) = 0.994. Thus the *mechanical* losses are only 0.6%.

14.3 Self-locking epicyclic trains

Some mechanisms, such as worm gearing and screw jacks, are self-locking; they cannot be driven by force or torque applied to the output member. This can be a valuable safety feature, ensuring, for example, that a jack will not 'run back' while you are under the car, or a hoist drop its load if the driving motor fails or the operator's hand slips on the rope. Obviously, however, self-locking can be a nuisance, since mechanisms are often required to overrun, particularly power transmissions when slowing down. Differential mechanisms can be self-locking, as we shall show by an example.

Consider a train with $\rho = 1.1$ and $E_0 = 0.9$, the arm as input and central gear 2 fixed. The table gives an efficiency of 0.45. This is not very satisfactory, but the train would run, and perhaps be suitable for the low speed, and hence low power, of a hoist. Now if we apply the load to central gear 1, so that the arm becomes the output member (the running back condition), the efficiency is given by the table as -0.1. This means that the arm, instead of giving out power, would actually absorb it and act as an input, i.e. that (in the case of a hoist) the load would have to rise instead of falling. The general rule is that if a mechanism has positive efficiency when driven normally, but zero or negative efficiency when the drive is applied to what should be the output member, it is self-locking.

It can be seen, by examining the formulae in the table, that trains with *negative* basic ratios can never be self-locking.

14.4 Other differential mechanisms
As in Chapter 13, our reasoning applies equally well to mechanisms giving linear instead of rotary motion, such as pulley blocks and differential screw jacks. It must be remembered, however, that in some of these mechanisms the value of the basic efficiency depends on the direction of power flow. This is particularly so in screw mechanisms.

Exercises 14
14.1. Gear-trains are required for transmission ratios of (*a*) 2 : 1, (*b*) – 3 : 1, (*c*) 1.1 : 1. Sketch the trains that you consider most efficient for each case and calculate their efficiencies, using only ordinary spur gears (external or internal).

14.2. A load torque of 100 N m is to be driven at 1000 revs/min by a motor with a speed of 900 revs/min, using a smaller motor and a differential. Design the system, ensuring if possible that neither motor acts as a brake, and calculate the power loss in kW in the differential.

14.3. A block is to be raised by applying a horizontal force to a wedge, as in Fig. 14.6. The coefficient of friction at all sliding surfaces is μ. Find the greatest angle θ for which the mechanism is self-locking.

Fig. 14.6.

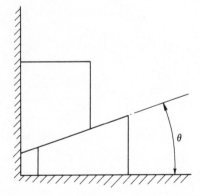

14.4. Can a hoist using only cords and pulleys ever be self-locking? Give reasons for your answer if it is no, or sketch a suitable design if it is yes.

14.5. Design a self-locking hoist using any mechanism you think suitable to lift a 1000 kg load with a 30 kg pull, taking losses into account.

14.6. This chapter explains in detail the derivation of the result in the table of efficiencies (Fig. 14.3) for an epicyclic train with shaft 1 fixed, shaft 2 as input and a positive basic ratio. Derive the result for the same case but with a negative basic ratio; and then derive the result for the first case in the table.

SUGGESTIONS FOR FURTHER READING

Descriptions of a great variety of useful mechanisms, with examples of their industrial application, can be found in *Mechanisms, linkages and mechanical controls* (N. P. Chironis, London: McGraw-Hill, 1965). For further and more serious study one should begin with *Kinematic synthesis of linkages* (R. S. Hartenberg & J. Denavit, London: McGraw-Hill, 1964), which includes spatial linkages as well as the plane mechanisms to which the present book is limited. From then on it is a matter of specialisation. The standard work on advanced graphical techniques, essential for the reader who wishes to undertake serious design work, is *Applied kinematics* (K. Hain, London: McGraw-Hill, 1964). But Hain's book is almost entirely about linkages; for information on cam mechanisms there are *Cams: design, dynamics and accuracy* (H. A. Rothbart, New York: Wiley, 1956), *Cam design and manufacture* (P. W. Jensen, New York: Industrial Press, 1965) and *The design of cam mechanisms and linkages* (S. Molian, London: Constable, 1968). A more advanced knowledge of theoretical kinematics can be found in *Motion geometry of mechanisms* (E. A. Dijksman, Cambridge University Press, 1976), *Kinematic geometry of mechanisms* (K. H. Hunt, Oxford University Press, 1978) and *Theoretical kinematics* (O. Bottema & B. Roth, North Holland, 1979). An emphasis on analytical methods, with many examples of computer programs, is found in *Kinematics and linkage design* (C. H. Suh & C. W. Radcliffe, Wiley, 1978). The journal *Mechanism and machine theory* (Oxford: Pergamon Press), which appears quarterly, is essential reading for anyone with a serious interest in this field.

INDEX